Problem Set

to accompany

BEDFORD•FOWLER ENGINEERING MECHANICS

Statics and Dynamics

Second Edition

by

Wallace Fowler
University of Texas, Austin

An imprint of Addison Wesley Longman, Inc.

Menlo Park, California • Reading, Massachusetts • New York • Harlow, England
Berkeley, California • Don Mills, Ontario • Sydney • Bonn • Amsterdam • Tokyo • Mexico City

Problem Set to accompany Bedford/Fowler's ENGINEERING MECHANICS: Statics and Dynamics, Second Edition

ISBN 0-201-36192-2

4 5 6 7 8 9 10—VG—01 00 99

Addison Wesley Longman, Inc.
2725 Sand Hill Road
Menlo Park, CA 94025

Contents

Statics

Chapter 1: Introduction

1.1ps The crawler that originally moved the Saturn V booster and now moves the space shuttle to the launch pad weighs 4.90×106 lb at sea level. What would the crawler weigh on the surface of the Moon? (The acceleration due to gravity on the surface of the Moon is 5.32 ft/s^2.)

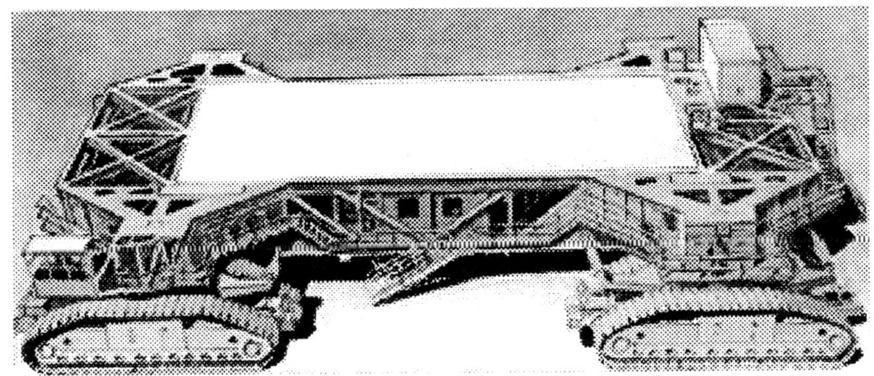

Figure P1.19, page 13

1.2ps The dimensions of the Boeing 777-200 aircraft are: length = 209 ft 1 in., wingspan = 199 ft 11 in., height (bottom of wheels to top of fin) = 60 ft 6 in., and interior cabin diameter = 19 ft 3 in. Express these dimensions in meters.

1.3ps In *The Weather Book* by Jack Williams, it is stated that an upward wind speed of 100 mph (miles/hr) is required for hailstones of 3-in. diameter to develop. Change these dimensions to m/s and mm.

1.4ps Thunderstorm cloud tops can reach altitudes as large as 18,500 m. What is this altitude in feet?

1.5ps The world's first successful parachute jump, in France in 1797, was made from a balloon from a height of 680 meters. What was the height of the jump in ft?

1.6ps The Santa Fe Railroad bought and operated 65 of the 2,900-class 4-8-4 steam locomotive, the largest non-articulated steam locomotive ever built. It had a mass of 436,818 kg. What was the weight of the locomotive in pounds?

1.7ps The radius of the Earth is 3,960 miles. What is the circumference of the Earth in km and in miles?

1.8ps Dick Rutan and Jeana Yeager flew the Voyager aircraft around the world without refueling. They flew a distance of 40,212.139 km in 9 days, 3 minutes and 44 seconds. What distance did they fly in miles? Determine their average speed (the distance flown divided by the time required) in mph, km/hr, ft/s, and knots (nautical miles per hour).

Chapter 2: Vectors

2.1ps The cables supporting the storage tank are shortened, decreasing the 20° angles to 30°. Determine the increases in the magnitudes of $\mathbf{F}_A$ and $\mathbf{F}_B$ caused by shortening the cables.

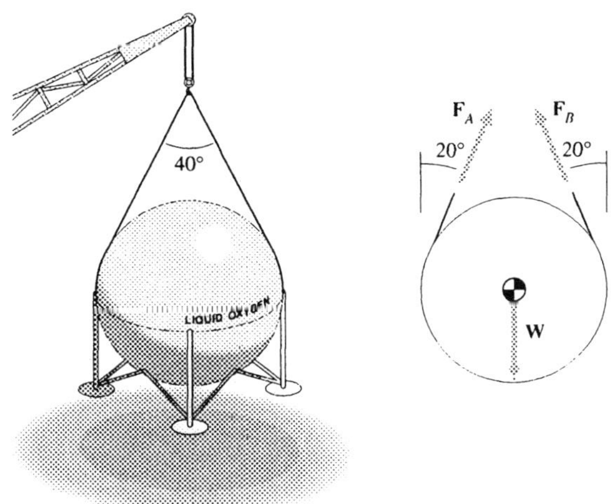

2.2ps A force is given in terms of its components by $\mathbf{F} = 14\mathbf{i} - 9\mathbf{j}$ (kN). What is its magnitude?
Strategy: The magnitude of a vector is given in terms of its components by Eq. (2.8).

2.3ps Write the position vectors from the point of contact of the rear wheel with the ground to points *A*, *B*, and *C* in terms of their components.

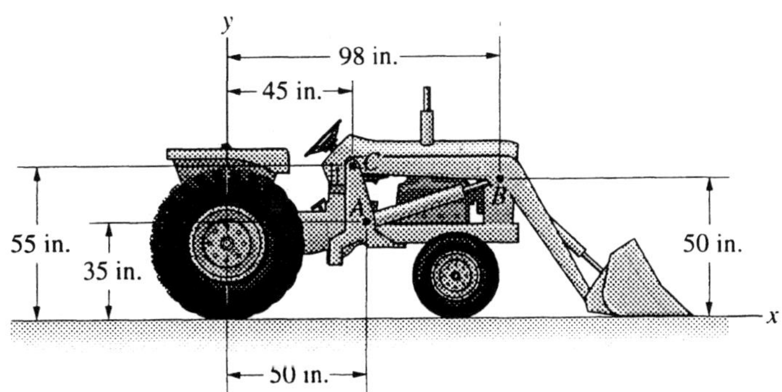

2.4ps The vector $\mathbf{r}_{OA} = 400\mathbf{i} + 800\mathbf{j}$ (m), $|\mathbf{r}_{AB}| = 400$ m, and $|\mathbf{r}_{OA} + \mathbf{r}_{AB}| = 1{,}200$ m. The surveyor who planned the road along the path *OAB* is also considering the alternative path going in a straight line from *O* to *B*. Determine the length and cost of each path if the road along path *OAB* costs $500,000 per km and the road along *OB* costs $600,000 per km.

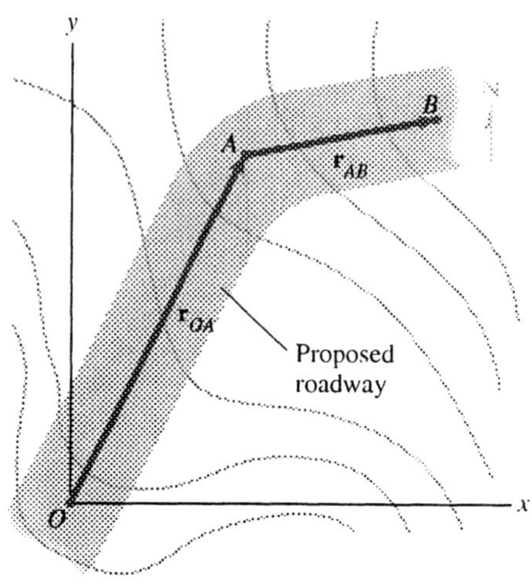

2.5ps A surveyor determines that the length of the line *OA* is 1,500 m and the length of line *OB* is 2,000 m. She then moves her transit from *O* 500 meters up along the *y* axis to a new origin *O′* and takes new sightings of points *A* and *B*.

(a) Express the position vector from her new position *O′* to point *A* in terms of scalar components.

(b) What is the angle between the position vector from *O′* to point *A* and the position vector from *O′* to point *B*?

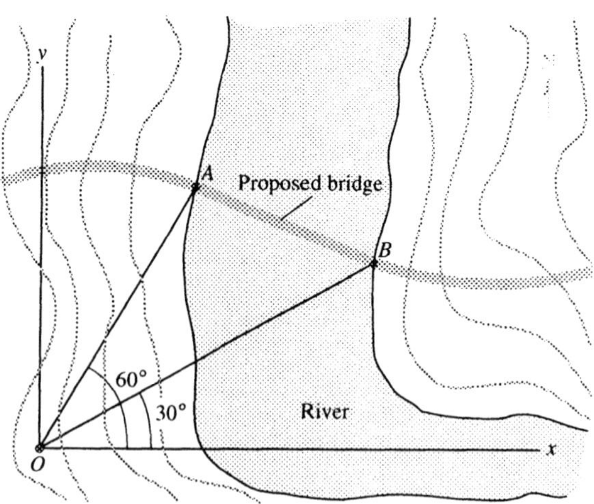

2.6ps The space shuttle performs simultaneous radar measurements of the satellites *A* and *B*. The onboard computer reports the positions to be $\mathbf{r}_A = 0.1\mathbf{i} - 0.2\mathbf{j}$ (km) and $\mathbf{r}_B = 0.5\mathbf{i} - 0.1\mathbf{j} - 0.4\mathbf{k}$ (km).

(a) What are the direction cosines of the vectors $\mathbf{r}_A$ and $\mathbf{r}_B$?
(b) Use the vectors **r**A and **r**B to determine the distance between the two satellites at the instant of the measurements.

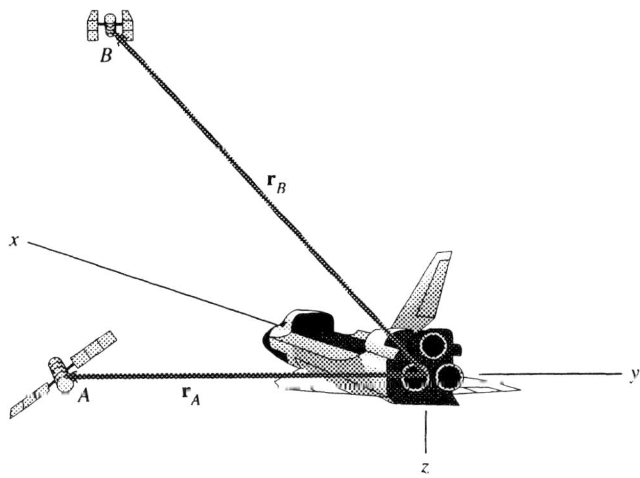

2.7ps The velocity of a satellite is $\mathbf{v} = 21{,}351\mathbf{i} + 12{,}051\mathbf{j} + 10{,}982\mathbf{k}$ (ft/s).
(a) What is the magnitude of its velocity?
(b) What are the angles between its velocity vector and the *x*, *y*, and *z* axes?

2.8ps Let *D* be the point of the pre-Columbian structure that is farthest from point *A*.
(a) Express the position vector $\mathbf{r}_{AD}$ from *A* to *D* in terms of its components.
(b) What are the direction cosines of $\mathbf{r}_{AD}$?

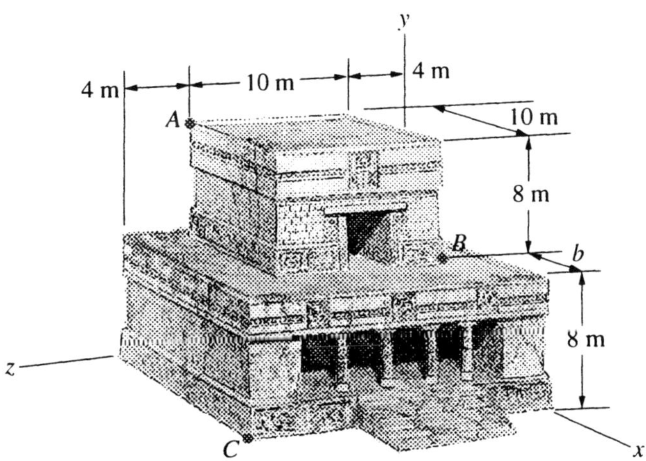

2.9ps The tension in the cable is 1,200 lb, so the cables *AB* and *AC* each exert a 1,200-lb pull on the bracket at *A*. What is the sum of the *y* components of the two forces on the bracket?

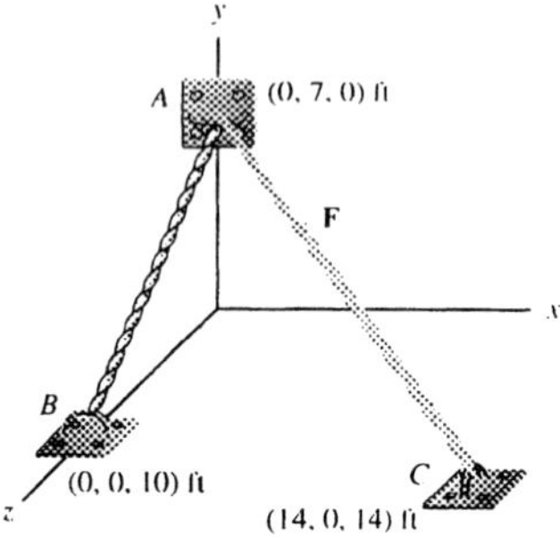

2.10ps Use the dot product to determine the perpendicular distance from point *C* to the line *AB*.

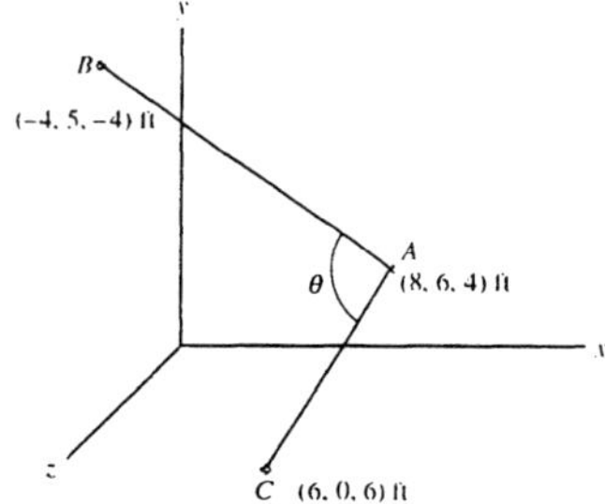

2.11ps Use the dot product to determine the perpendicular distance from the point *B* in Problem 2.10ps to the line *AC*.

2.12ps Use the dot product to determine the perpendicular distance from the ship *A* to the line *OB*.

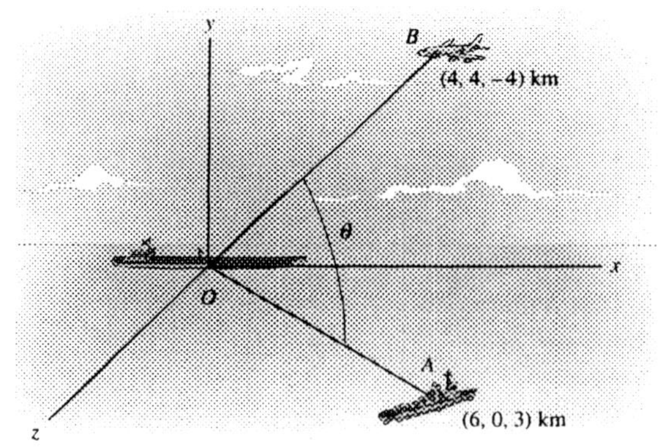

2.13ps Let $\mathbf{r}_{OB}$ be the position vector from the origin to point *B*. Determine the cross product $\mathbf{r}_{OB} \times \mathbf{F}$.

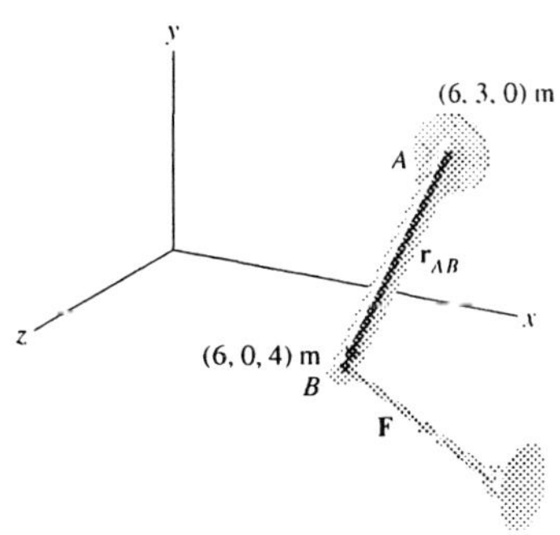

2.14ps A 30-N force F lies along the line from *A* to *B*. Determine the cross product $\mathbf{r}_{CA} \times \mathbf{F}$.

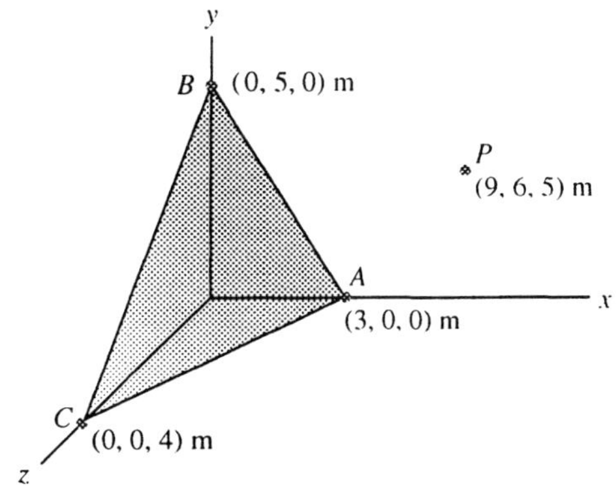

2.15ps The magnitude of the force $\boldsymbol{A}$ is 200 lb, and $\boldsymbol{A} + \boldsymbol{B} + \boldsymbol{C} = 0$. Denote the top point of the truss as point *P*. Determine the cross products $\mathbf{r}_{PA} \times A$ and $\mathbf{r}_{PB} \times B$.

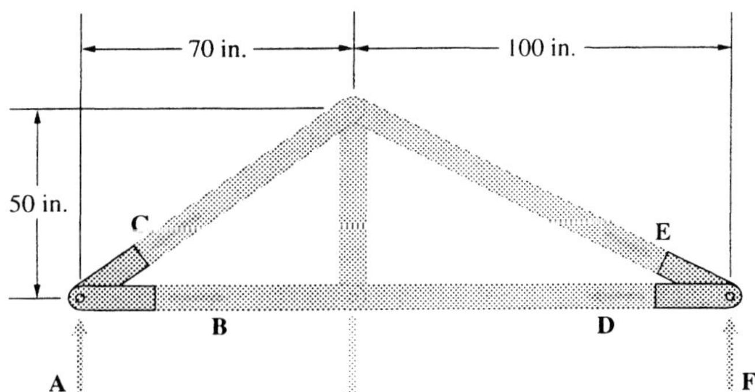

2.16ps The magnitude of the force vector $\mathbf{F}_A$ is 8 kN. Determine the component of $\mathbf{F}_A$ that is parallel to the line AB. (Your answer should be a vector.) Strategy: The component of a vector parallel to a line is given by Eq. (2.26).

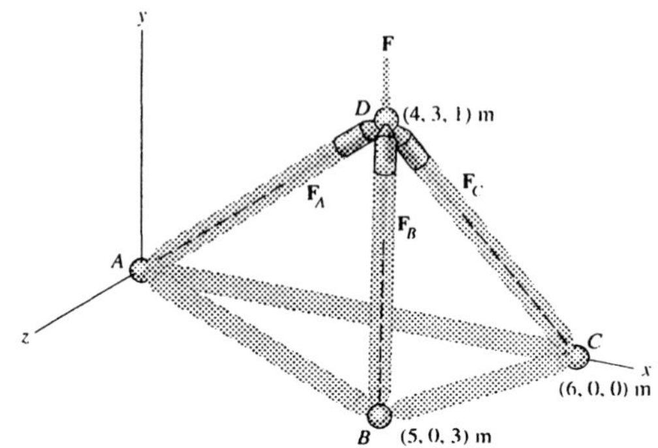

2.17ps The magnitude of the force vector $\mathbf{F}_B$ in Problem 2.16ps is 6 kN. Determine the component of $\mathbf{F}_B$ that is is parallel to the line AC.

2.18ps The magnitude of the force vector $\mathbf{F}_B$ in Problem 2.16ps is 6 kN. Determine the cross product $\mathbf{r}_{AD} \times \mathbf{F}_B$.

Chapter 3: Forces

3.1ps The magnitudes of the forces **B** and **C** are 400 N and 760 N, respectively. If the sums of the forces in the x and y directions are both zero, what are the magnitudes of the forces $\mathbf{A}_x$ and $\mathbf{A}_y$?

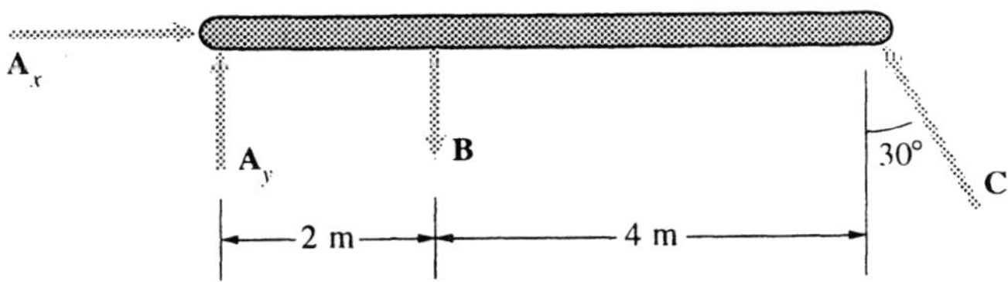

3.2ps For the beam shown in Problem 3.1ps, suppose that the magnitudes of the forces $\mathbf{A}_y$ and **C** are 1,000 N and 1,200 N, respectively. Assume that the sums of the forces in the x and y directions are both zero. What are the magnitudes of the forces $\mathbf{A}_x$ and **B**?

3.3ps Redesign the system of pulleys so that the force required to support the weight W is $W = F/7$. (You may add or subtract pulleys from the system.)

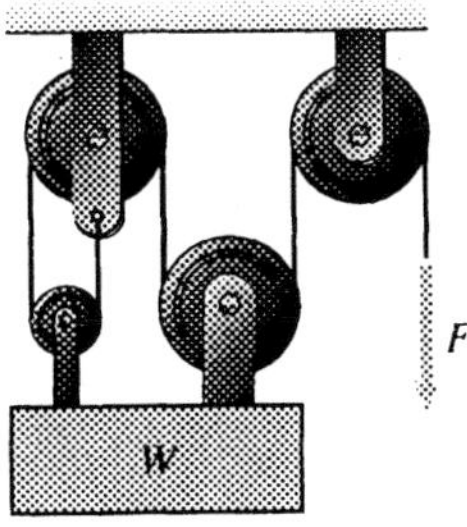

3.4ps The mass of the painting is 10 kg. Draw a graph of the tension in the wire as a function of α for $5° < \alpha < 45°$. Given that the wire will break at a tension of 400 N, recommend an angle α to be used in hanging the painting.

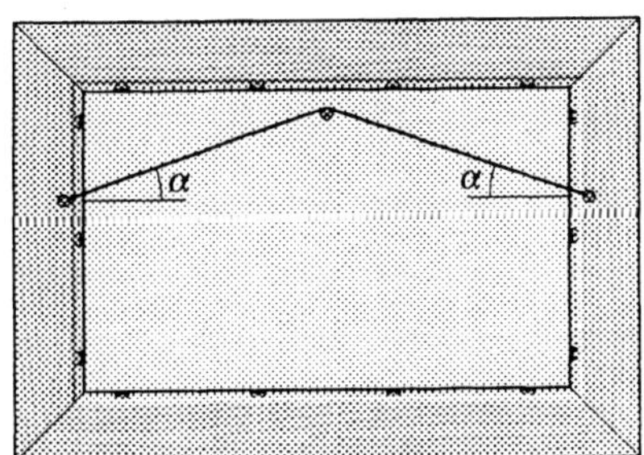

3.5ps If the construction worker on the moon (acceleration due to gravity = 5.32 ft/s^2) can comfortably exert a force of 8 lb on the cable, what is the earth weight of the heaviest crate she can comfortably hold in the position shown?

3.6ps What is the largest weight the system shown will support if the maximum tension in any cable must not be greater than 2,000 lb?

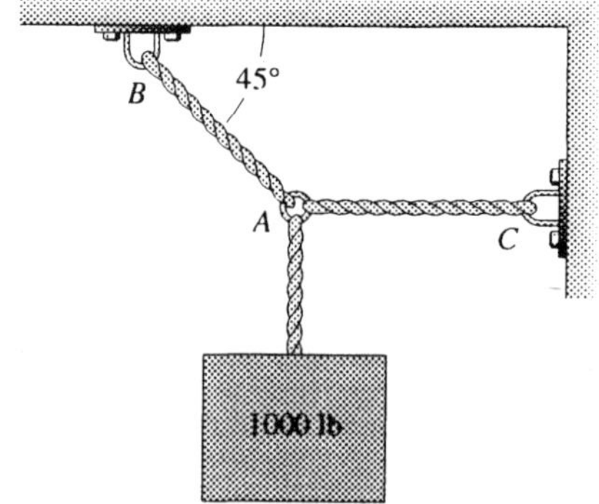

3.7ps Suppose that you wish to support the 1,000-lb weight in Problem 3.6ps in the position shown. You are free to move the supports *B* and *C*, changing the lengths of the cables appropriately. Determine the locations for *B* and *C* which minimize the tensions in cables *AB* and *AC*. What are the tensions?

3.8ps The mass of the traffic light is 140 kg. If the cables *AB* and *AC* can each safely support a tension of 8 kN, what distance above its present position could the light safely be raised by shortening cables *AB* and *AC*?

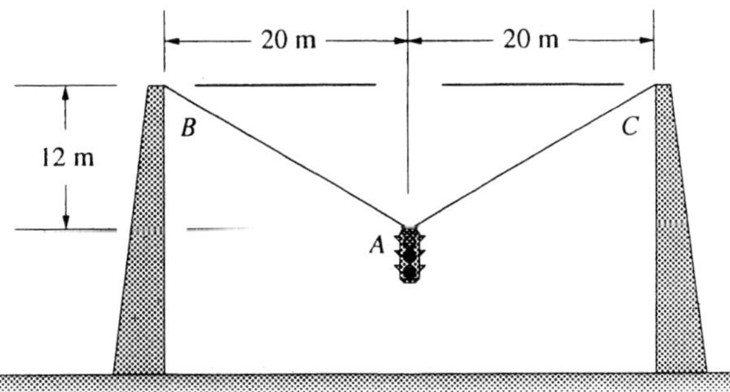

3.9ps The angle $\beta = 50°$ and the tension $T = 2$ kN. Draw graphs of the tensions in the cables *AB* and *AC* as functions of α for $-20° < \alpha < 90°$.

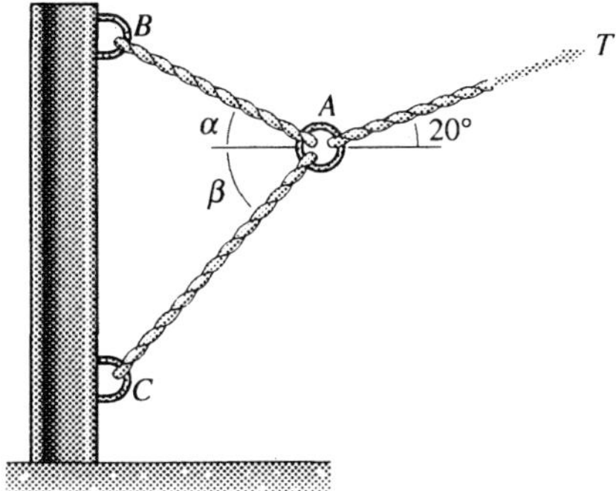

3.10ps The surfaces are smooth. Determine the angle α in the range $5° < \alpha < 90°$ for which the magnitude of the force exerted on the 50-lb cylinder by the right-hand surface is smallest. Explain why this configuration results in the smallest force.

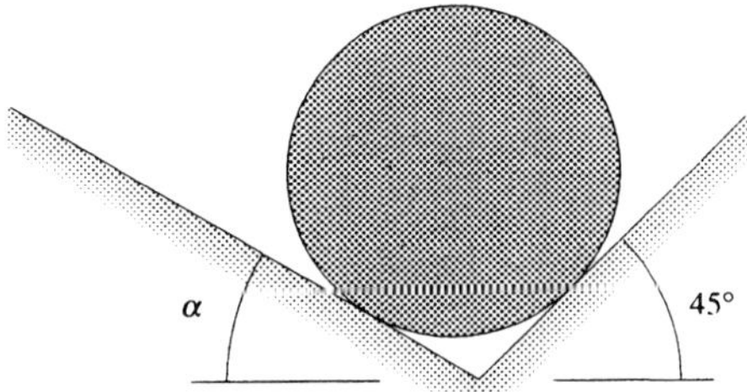

3.11ps For the system in Problem 3.10ps, determine the angle α in the range $5° < \alpha < 90°$ for which the magnitude of the force exerted on the 50-lb cylinder by the left-hand surface is smallest. Explain why this configuration results in the smallest force.

3.12ps If the middle weight of the system is increased to $1.5W$ and the system is in equilibrium, what are the coordinates of point A?

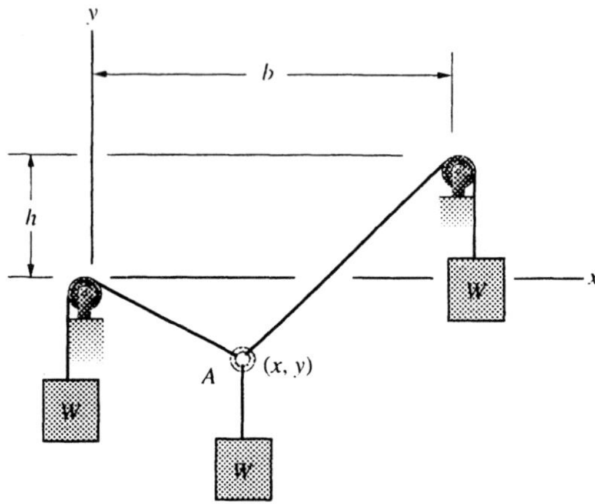

3.13ps If the left-hand weight of the system shown in Problem 3.12ps is increased to $2W$ and the system is in equilibrium, what are the coordinates of point A?

3.14ps The sum of the downward forces exerted at A and B by the patient's leg is 150 N. What horizontal force is exerted on the patient's leg at A by the Russel's traction apparatus?

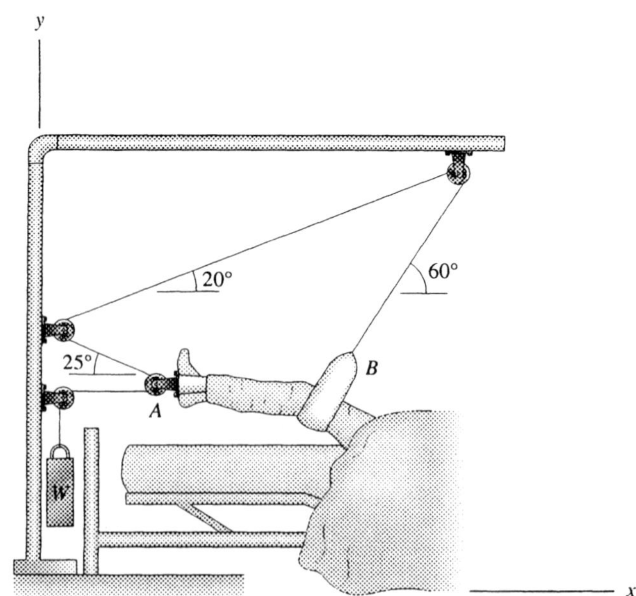

3.15ps The mass of block A is 42 kg, the mass of block B is 50 kg, and the surfaces are smooth. If the 45° angle is changed to 30°, what force F is necessary to hold the blocks in equilibrium?

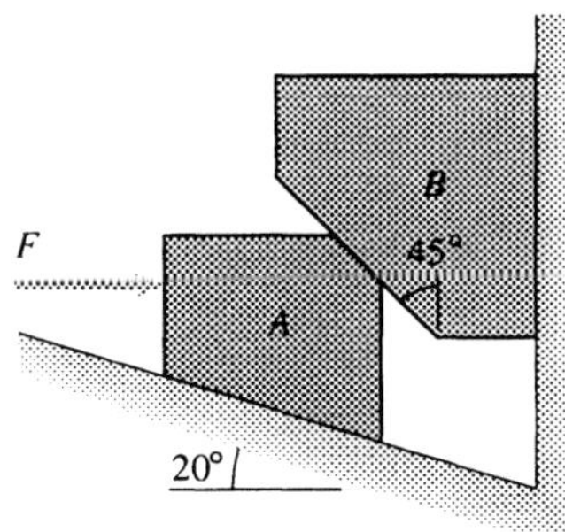

Chapter 4: Systems of Forces and Moments

4.1ps The force $\mathbf{F} = 47\mathbf{i} - 33\mathbf{j}$ (N) acts at the point (4,7) m. What is the moment of **F** about the point (1, 3) m?

4.2ps The force $\mathbf{F} = 47\mathbf{i} - 33\mathbf{j}$ (N) acts at the point (4,7) m. The moment of **F** about a point *P* is $\mathbf{M} = 33.5\mathbf{k}$ (N-m). What is the perpendicular distance from *P* to the line of action of **F**?

4.3ps (a) If the 20° angle is changed to 45°, what is the moment of the 80-lb force about *P*? (b) What angle should the 20° angle be changed to so that the counterclockwise moment of the 80-lb force about *P* is a maximum?

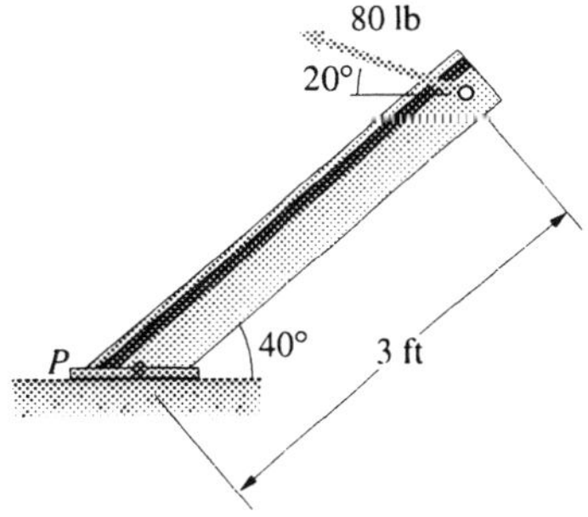

4.4ps Draw a graph of the moment of the 20-N force *F* about *P* as a function of the angle α for 0 < α < 90°.

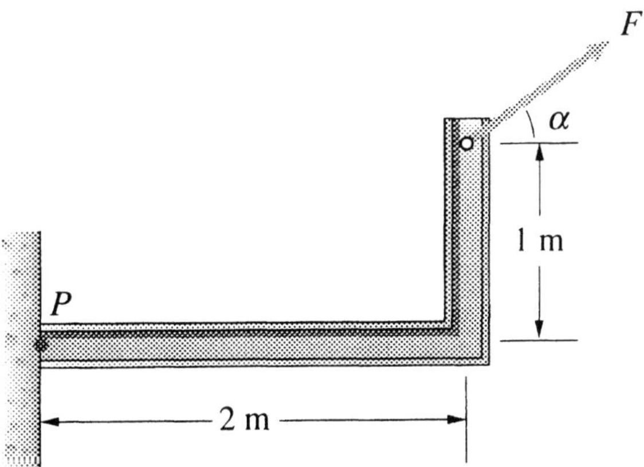

4.5ps The 150-lb downward force on the left of the lug wrench is caused by a 150-lb male student standing on it. At the same time, his female companion pulls on the right side of the wrench with a force *F* = 70 lb. As an alternative tactic, the 105-lb female could stand on the left side of the wrench while the male exerts a force $F = 120$ lb on the right side. Which way should they do it to exert the greatest total moment on the lug nut, and what is the total moment exerted in each case?

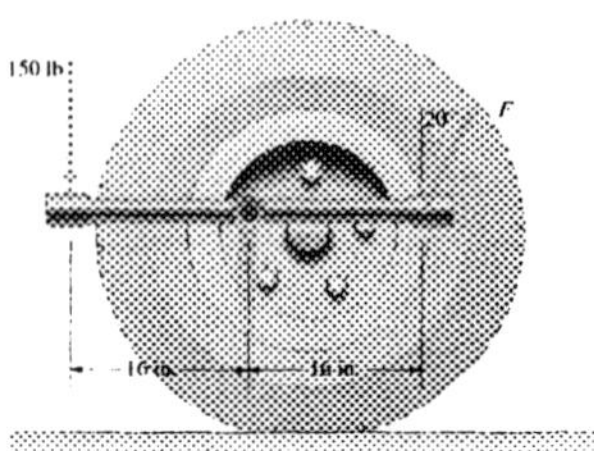

4.6ps The force F_2 (100 N) can be moved along the beam while F_1 (50 N) remains in the position shown. Let x_P be the distance from A to the point of application of F_2. Draw a graph of the sum of the moments of the two forces about A for $0 < x_P < 6$ m.

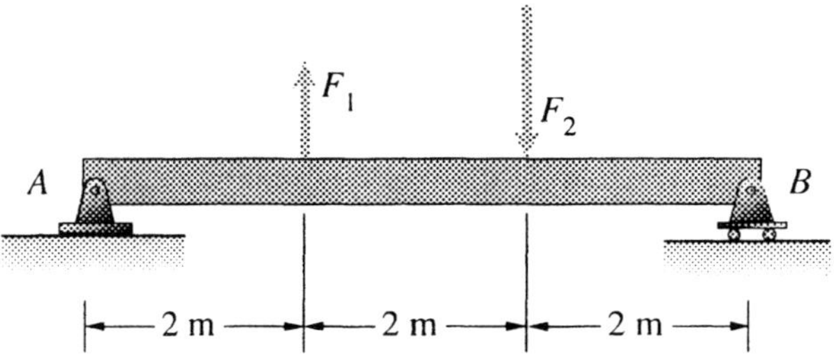

4.7ps What force acting at a point along the line AB will cause the sum of the forces on the object to be zero and will simultaneously cause the sum of the moments about A to be zero? Where is the point of application of the force located on the line AB?

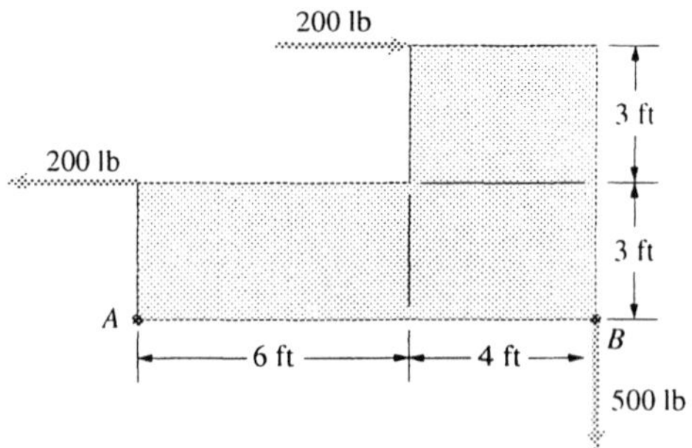

4.8ps The space shuttle's attitude thrusters exert two forces of magnitude $F = 1{,}740$ lb. If you represent these two forces by a single equivalent force, what is the force and what is the perpendicular distance from G to its line of action?

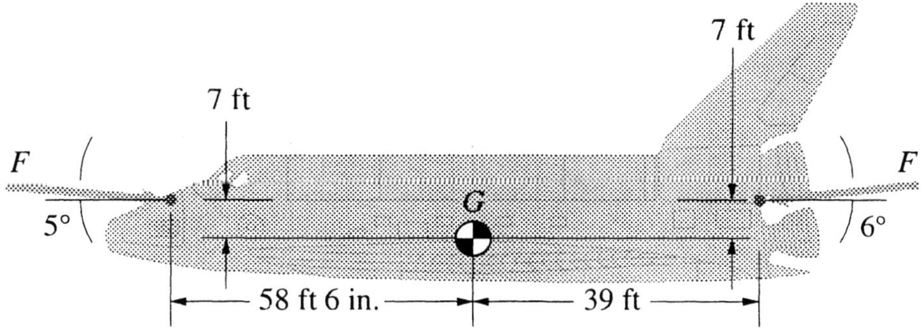

4.9ps The man slowly flexes his arm over the range $0 < \alpha < 60°$. Draw a graph of the moment he must exert as a function of α..

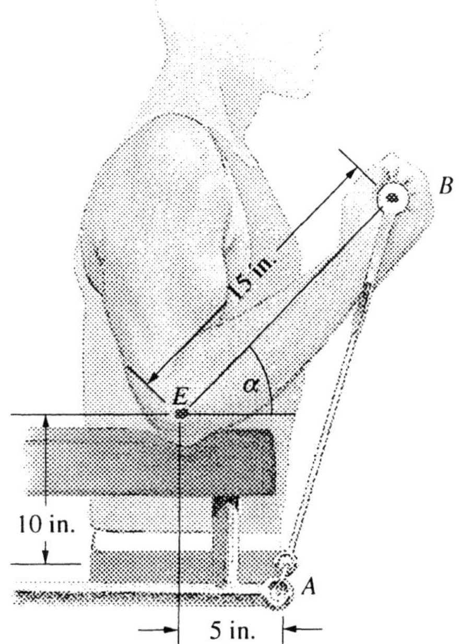

4.10ps The force $\mathbf{F} = 35\mathbf{i} - 22\mathbf{j}$ (N) is applied at the point (20, 10) m. Use the cross product to determine the moment of **F** about the point (1, 2) m.

4.11ps The force $\mathbf{F} = 15\mathbf{i} + 20\mathbf{j} - 19\mathbf{k}$ (N) is applied at the point (25 ,5 ,7) m. Use the cross product to determine the moment of **F** about the point (1, 2, 10) m.

4.12ps The moment of a force $\mathbf{F} = 5\mathbf{i} + 10\mathbf{j} - 20\mathbf{k}$ (N) about the point (5, 10, 10) m is $\mathbf{M} = 40\mathbf{i} + 60\mathbf{j} + 40\mathbf{k}$ (N-m). Determine the coordinates of the point P where the line of action of **F** crosses the x-y plane.

4.13ps What tensions in the cables AB and AC would produce a horizontal component of force of $100\mathbf{i} + 240\mathbf{k}$ (lb) on the tree?

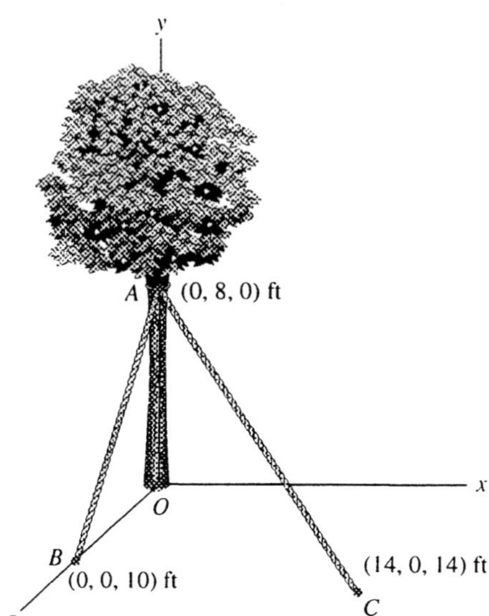

4.14ps The distance $b = 2$ m. Draw a graph of the moment exerted by the couple if the 35° angle is varied over the range from 0 to 180°.

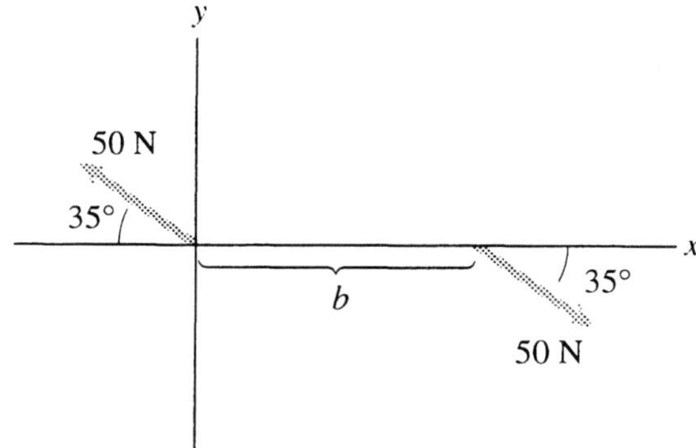

4.15ps The airplane's pilot aborts a takeoff and engages the thrust reverser on each engine. The resulting thrusts are:

engine 1:	–50 kN
engine 2:	–45 kN
engine 3:	–50 kN
engine 4:	–55 kN

If you represent these four braking forces by a single equivalent force **F**, what is **F** and where does its line of action intersect the *y* axis?

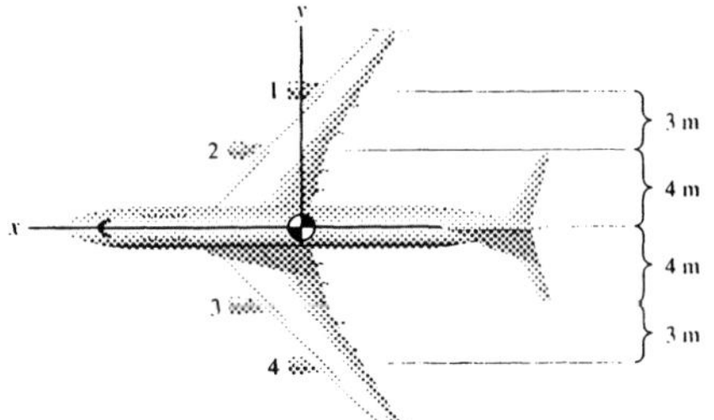

4.16ps If you represent the two forces acting on the beam by a force **F** acting at the midpoint of the beam and a couple **M**, what are **F** and **M**?

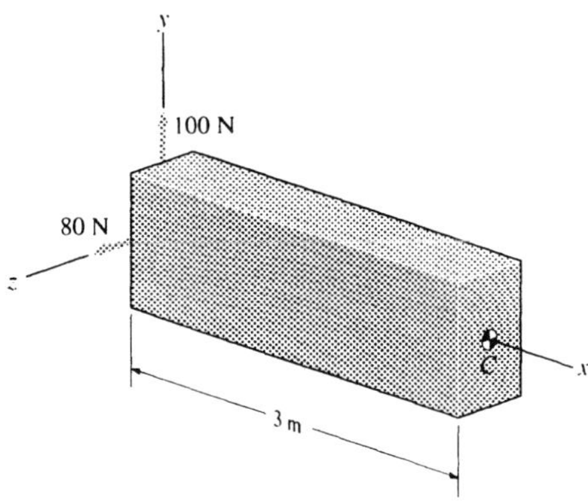

4.17ps The forces acting on the block are

$$\mathbf{F}_A = -5\mathbf{i} + 15\mathbf{j} + 10\mathbf{k} \text{ (lb)},$$
$$\mathbf{F}_B = 10\mathbf{i} + 15\mathbf{j} + 2\mathbf{k} \text{ (lb)}.$$

If you represent these forces by a force **F** acting at the origin of the coordinate system and a couple **M**, what are **F** and **M**?

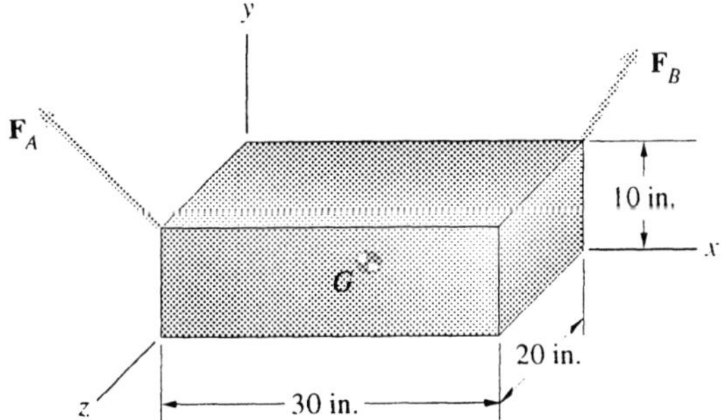

4.18ps The engine above the fuselage of the DC-10 exerts a thrust $T_0 = 16$ kip, and each of the engines under the wings exerts a thrust $T_U = 12$ kip. The dimensions are $h = 8$ ft, $c = 12$ ft, and $b = 16$ ft. Suppose that the engine under the wing to the pilot's left experiences flame-out (completely loses thrust). If you represent the two remaining thrust forces by a force **F** acting at the origin and a couple **M**, what are **F** and **M**?

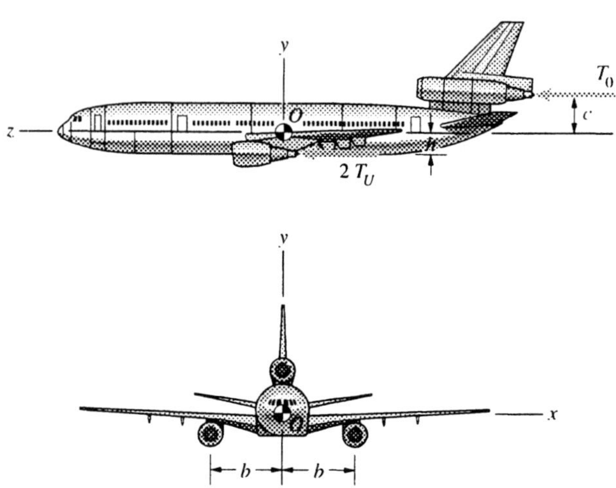

4.19ps Suppose that the engine above the fuselage of the DC-10 described in Problem 4.18ps experiences flame-out. If you represent the two remaining thrust forces by a force **F** acting at the origin and a couple **M**, what are **F** and **M**?

4.20ps The force **F** and couple **M** in system 1 are

$$\mathbf{F} = 10\mathbf{i} + 15\mathbf{j} + 5\mathbf{k}$$
$$\mathbf{M} = 8\mathbf{i} + 10\mathbf{j} - 8\mathbf{k} \text{ (ft-lb).}$$

Suppose you want to represent system 1 by a wrench (system 2). Determine the couple $\mathbf{M}_P$ and the coordinates x and z where the line of action of the force intersects the x-z plane.

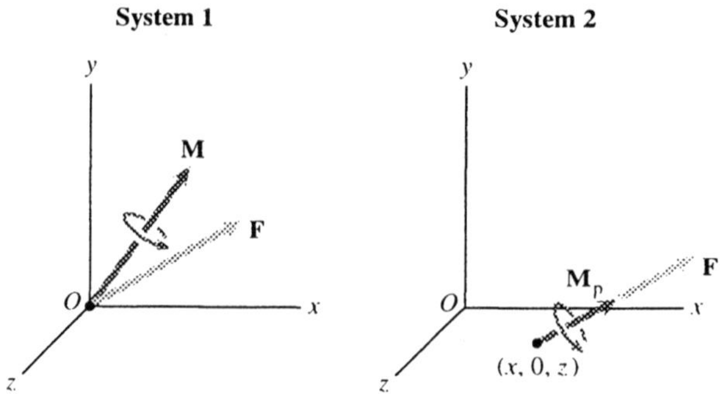

4.21ps Let a point B be located at (300, 200, 0) mm. Determine the moment of the 200-N force about B. (Express your answer as a vector.)

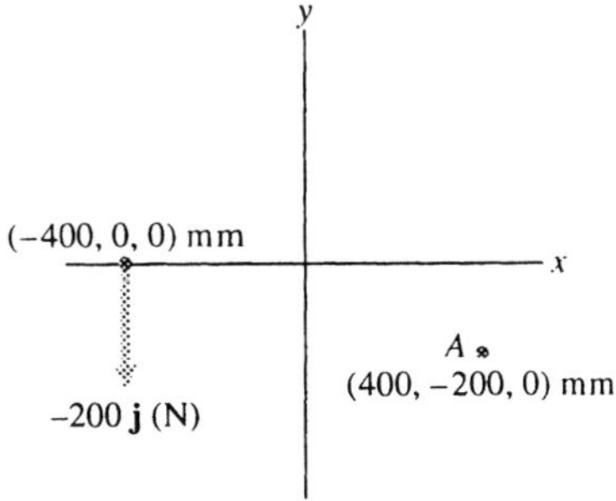

4.22ps Suppose that an additional force $\mathbf{C} = 200\mathbf{i} + 250\mathbf{j}$ (N) is applied at the "corner" of the L-shaped beam. If the sum of the forces on the beam is zero and the sum of the moments about B is zero, what are the forces A_x, A_y, and B?

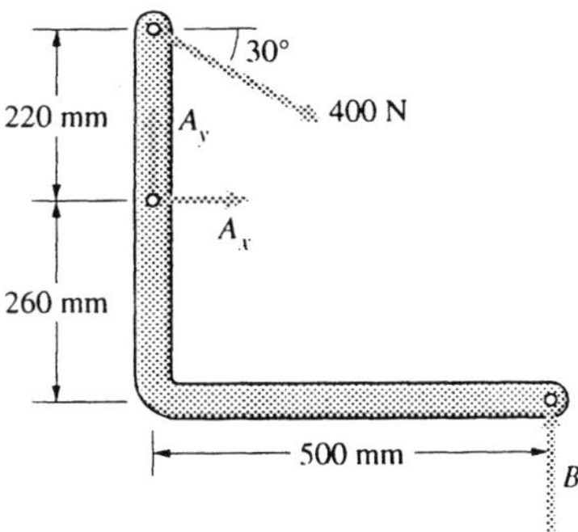

4.23ps The 20-kg mass is suspended by cables attached to three vertical 2-m posts. Point A is at (0, 1.2, 0) m. Suppose that the rightmost post is moved so that point D is at (2, 2, 2) m and the cable AD is lengthened accordingly. Determine the moment about the base E due to the force exerted at B by the cable AB.

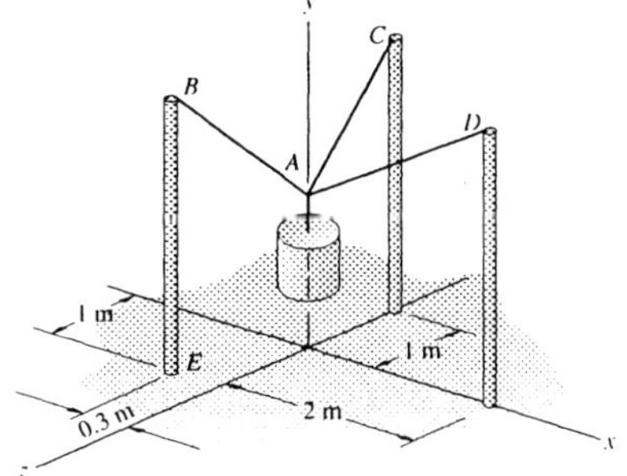

4.24ps The two force systems shown are equivalent. If you represent system 2 by an equivalent single force **F**, what is **F** and where does its line of action intersect the x axis?

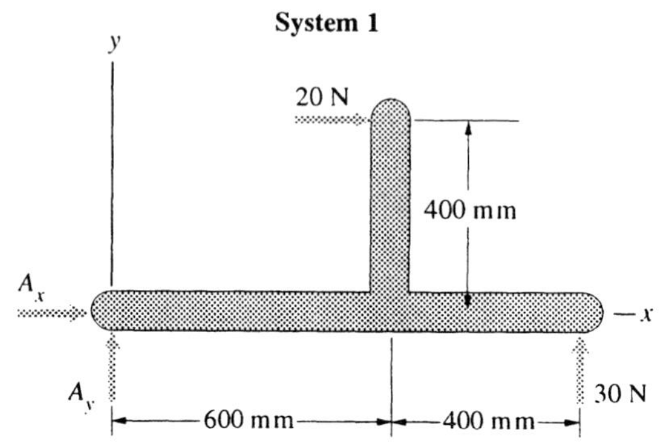

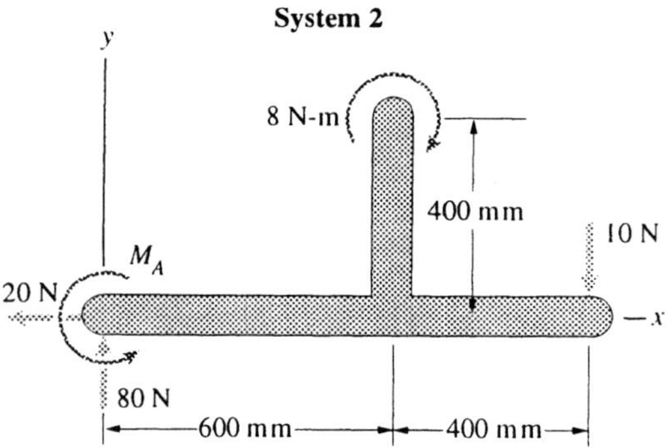

4.25ps If you represent system 1 by an equivalent system consisting of a force $\mathbf{F}_B$ acting at the base of the pipe assembly (0, 0, 6) in. and a couple $\mathbf{M}_B$, what are $\mathbf{F}_B$ and $\mathbf{M}_B$?

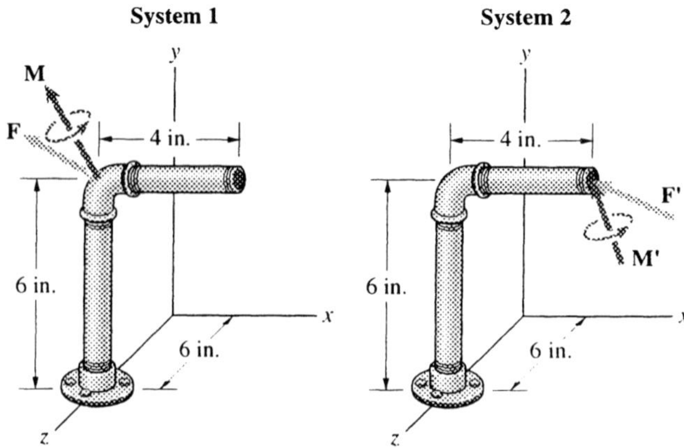

Chapter 5: Objects in Equilibrium

5.1ps Draw a free-body diagram of each of the following objects:
(a) A ladder resting on a rough floor and leaning on a smooth wall.
(b) A ladder resting on a rough floor and leaning on a rough wall.
(c) A straight beam of negligible weight extending 4 m out from the side of a building. An 80-kg object is suspended by a cable from the end of the beam.
(d) A 3,000-lb car parked with its front end facing uphill on a street that is inclined 30° from the horizontal. The parking break locks the rear wheels only; the front wheels are free to rotate.

5.2ps The airplane's weight is $W = 2{,}400$ lb. Its engine is not operating, so the thrust $T = 0$. A mechanic pulls downward on the horizontal stabilizer at a point 11 ft behind the main landing gear (B). Draw the free-body diagram of the airplane and determine the force the mechanic must exert to reduce the vertical force exerted on the nose wheel at A to 500 lb.

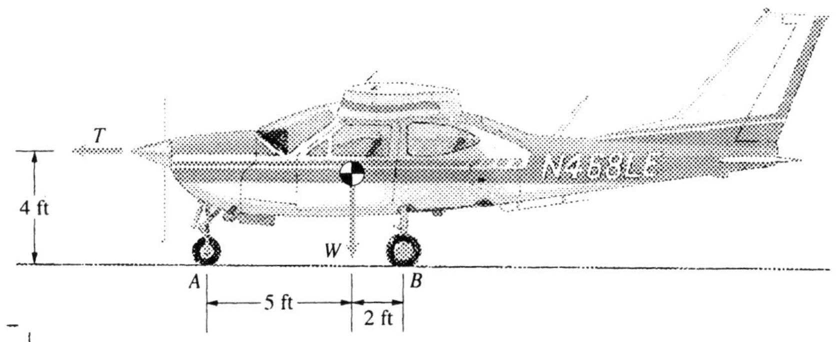

5.3ps The figure shows the top view of a fan of weight W. The fan is supported by small pads at the ends of the three legs of length b. The fan's thrust T is exerted at a height h above the floor. The design engineer realizes he has made a mistake pointing the fan relative to the legs in the direction shown, and rotates it 60°. Determine the ratio of the largest thrust T the fan can exert without tipping over with the fan in the new direction to the largest value with the fan in the direction shown.

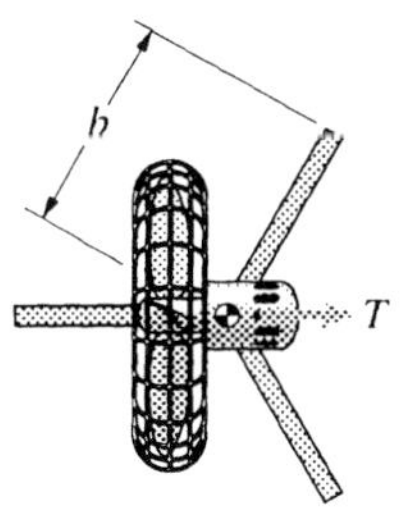

5.4ps The weight of the gymnast's left arm and the weight of his body *it not including his arms* are shown. The dimensions are $a = b = 9$ in. and $c = 13$ in.

(a) Treat his shoulder as a built-in support and determine the magnitudes of the reactions at his shoulder.

(b) For a second gymnast the arm and body weights are the same but the dimensions are $a = 9$ in., $b = 11$ in., and $c = 14$ in. Determine the magnitudes of the reactions at his shoulder and comment on the implications of having shorter or longer arms for this event.

5.5ps Assume that the car's parking brake locks all four wheels and that the friction force exerted on each wheel by the road is proportional to the normal component of force exerted on that wheel by the road. Determine the friction and normal forces on each wheel: (a) if $\alpha = 15°$; (b) if $\alpha = -15°$.

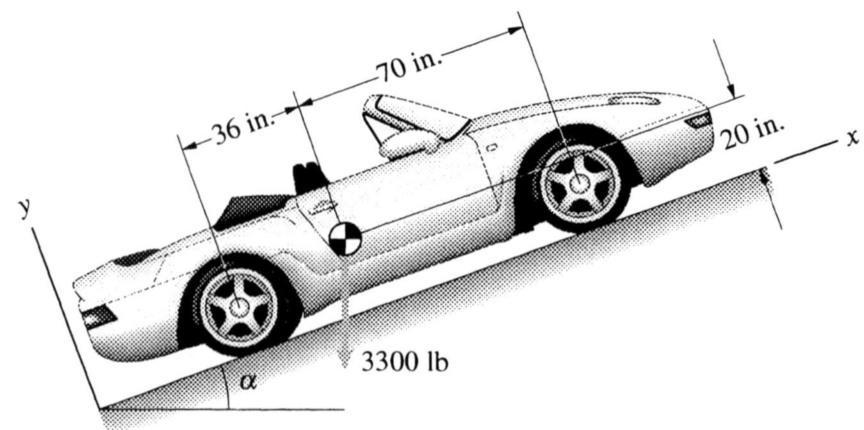

5.6ps Determine the effects on the reactions at the built-in support if the tension in cable *BC* is increased from 100 lb to 150 lb.

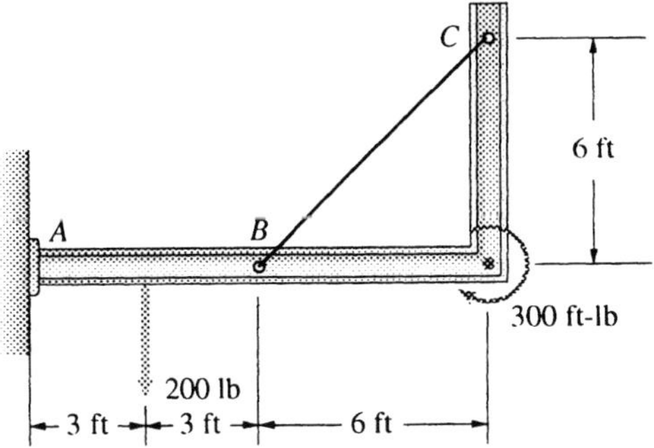

5.7ps For each of the cases shown, determine the effects on the reactions at C of increasing the tension in cable *AB* from 2 kN to 4 kN.

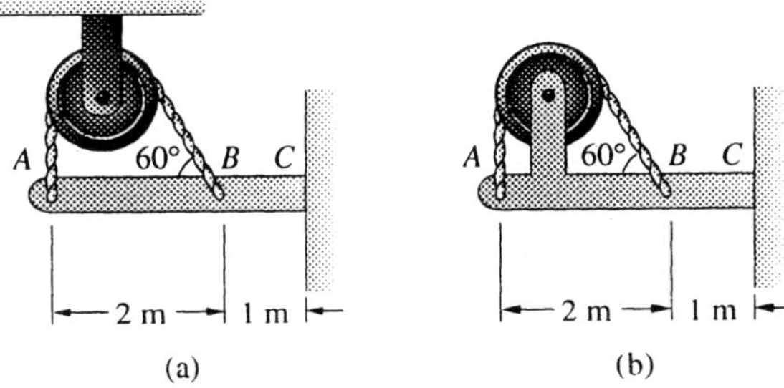

5.8ps Assume that the force *F* points upward instead of horizontally. Obtain an equation for the value of *F* necessary for the rectangular plate to be in equilibrium. (Your answer will be in terms of α, *W*, *b*, and *h*.)

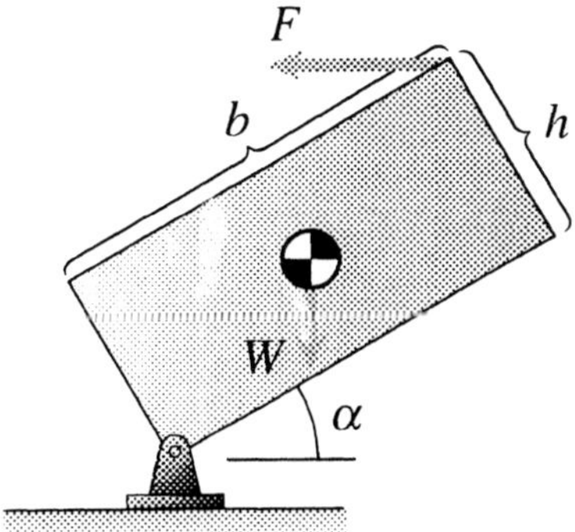

5.9ps The bar *AB* is of length L and weight W, and $\alpha = 30°$. Assume that the height of point C above the support at A is an arbitrary distance h instead of L. Draw a graph of the tension in string BC as a function of h for $0.6L < h < 2L$.

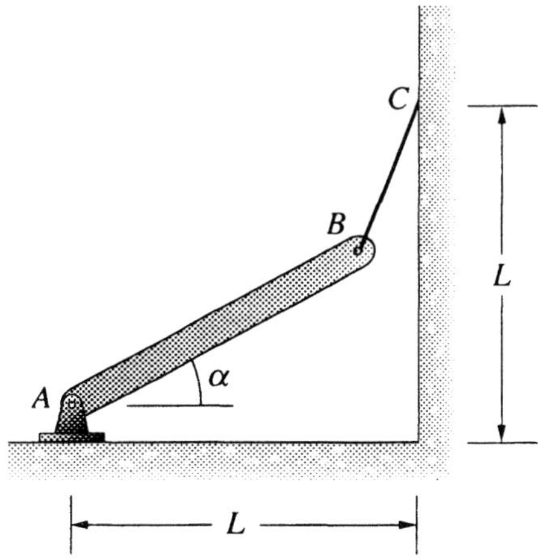

5.10ps The boom derrick slowly lowers the 15-kip load shown by changing the angle θ from 15° to 60°. The distances a and b are 15 ft and 2 ft, respectively. The lengths *BC* and *DE* are both 20 ft. Draw graphs of the tension in cable *AB* and the reaction at the pin support C as functions of θ.

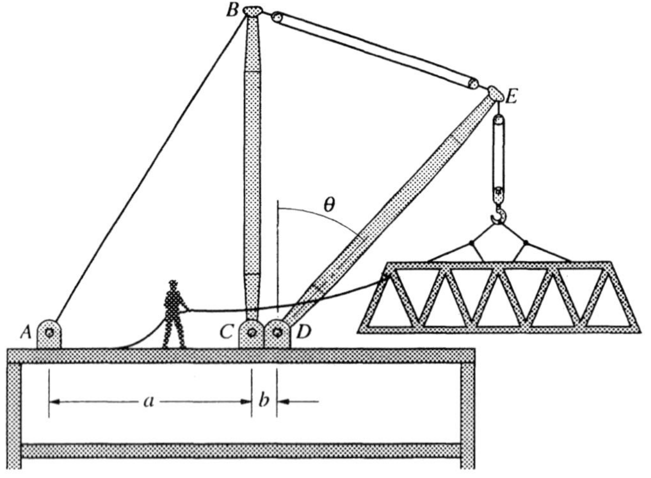

5.11ps The forces and couple on the bar are

$$\mathbf{F}_B = 4\mathbf{i} - 3\mathbf{j} + 5\mathbf{k} \text{ (kN)},$$
$$\mathbf{F}_C = 2\mathbf{i} + 4\mathbf{j} + 3\mathbf{k} \text{ (kN)},$$
$$\mathbf{M}_C = \mathbf{i} - 3\mathbf{j} + 2\mathbf{k} \text{ (kN - m)}.$$

(a) Draw the free-body diagram of the bar.
(b) Determine the reactions at the built-in support *A*.

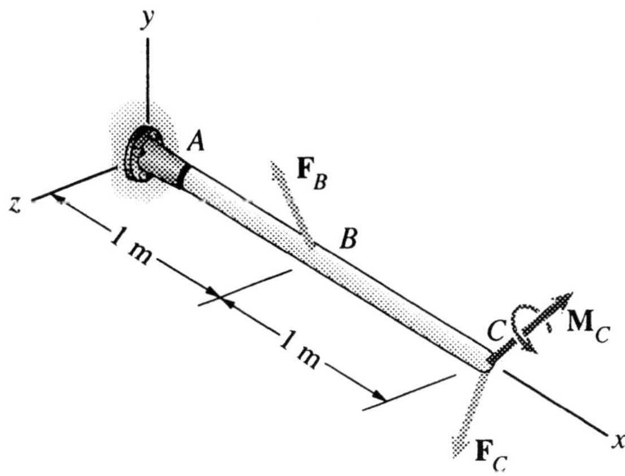

5.12ps Suppose that the attachment point *C* is moved to (4, 0, −2) m and the tension in cable *BC* is 25 kN.
(a) Draw the free-body diagram of the bar.
(b) Determine the reactions at the built-in support *A*.

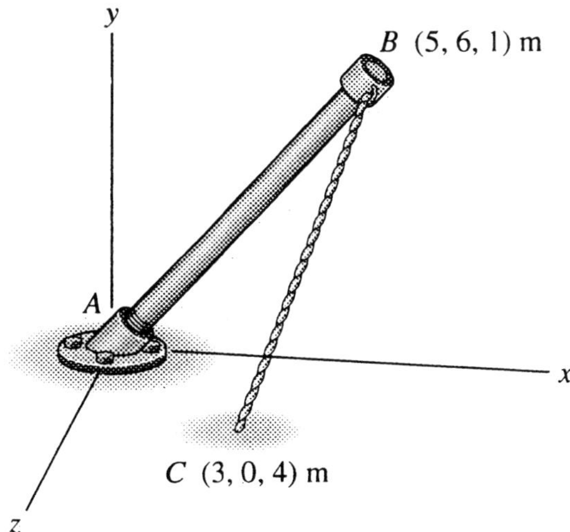

5.13ps Suppose that the collar *C* is moved along the bar until it is 6 feet from the wall and the cables are adjusted accordingly. The force $F = -1{,}000\mathbf{j}$ (lb).
(a) Draw the free-body diagram of the bar.
(b) Determine the tensions in the two cables and the reactions at the built-in support *A*.

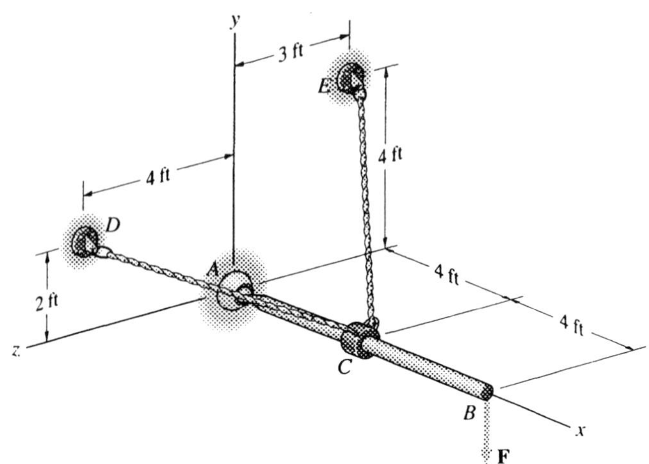

5.14ps Let the distance from the wall to the collar *C* in Problem 5.13ps be x_C. The force $\mathbf{F} = -1{,}000\mathbf{j}$ (lb). Draw graphs of the tensions in cables *CD* and *CE* for $0.5 < x_C < 8$ ft.

5.15ps The vertical force $F = 800$ lb. Draw graphs of the reactions at *A* and *B* as functions of *b* for $0 < b < 2.5$ ft.

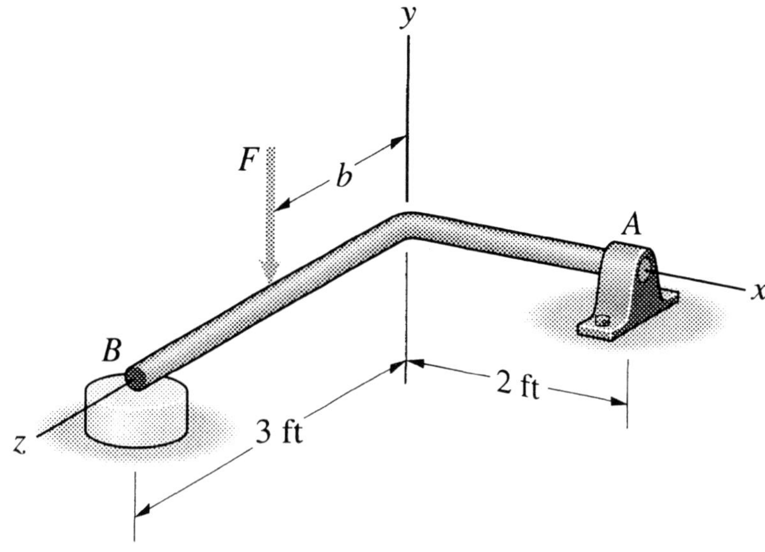

5.16ps As an engineering intern you are asked to investigate the effects of the locations of hinges on a typical door. Assume that the door shown weighs 40 lb, the hinges do not exert couples on the door, and the hinge at *B* does not exert a force parallel to the hinge axis. Let the vertical distance from the bottom of the door to hinge *A* and the equal vertical distance from the top of the door to hinge B be denoted by h. Draw graphs of the magnitudes of the reactions at A and B as functions of h for $0.5 < h < 3$ ft.

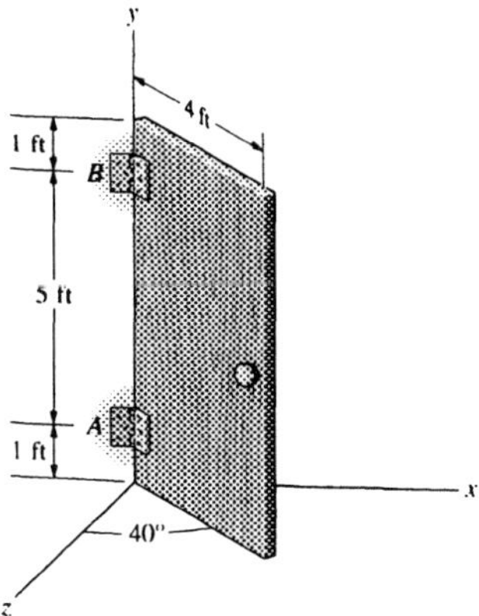

5.17ps Suppose that the 30° angle is changed to 20°. The horizontal bar is of negligible weight. Use the fact that the bar is a three-force member to determine the angle α necessary for equilibrium.

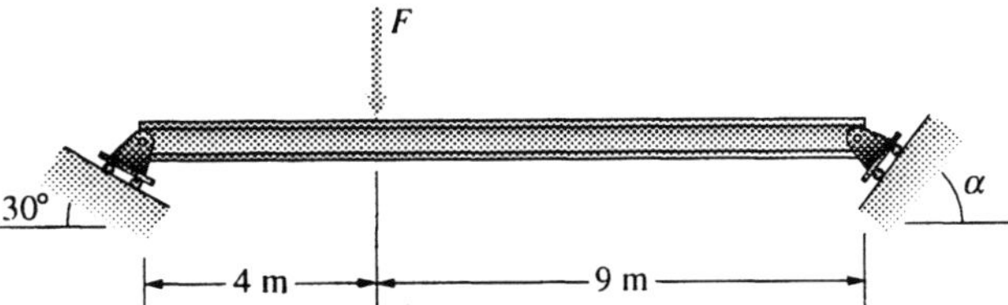

5.18ps The suspended load weighs 1,000 lb and the structure is of negligible weight. Let h be the vertical distance between the supports A and B. Use the fact that the structure is a three-force member to determine the reactions at A and B and draw graphs of them for $2 < h < 8$ ft.

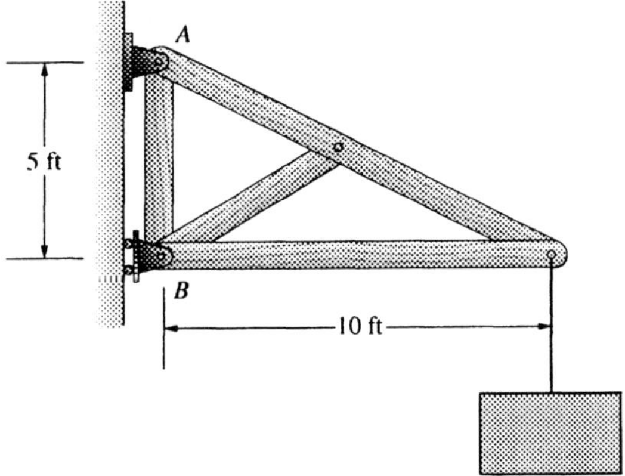

5.19ps Suppose that the height of the step is increased from 50 mm to 75 mm. The mass of the disk is 10 kg and the surfaces are rough. Determine the force F necessary to lift the disk off the floor.

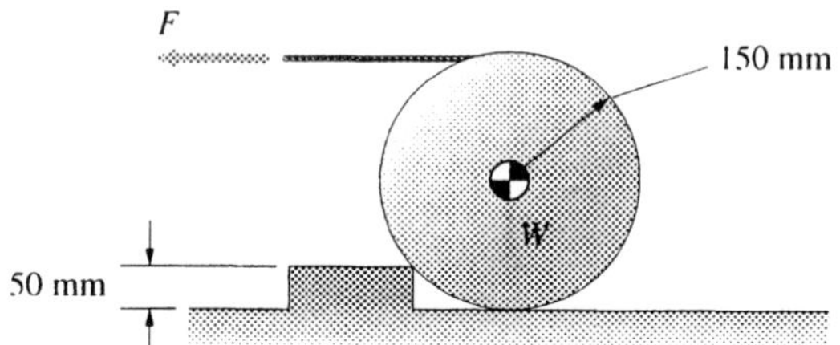

5.20ps For the disk described in Problem 5.19ps, determine the force F necessary to lift the disk off the floor as a function of step height h for $10 < h < 100$ mm.

5.21ps For the disk described in Problem 5.19ps, explore whether a large enough force F will lift the disk off the floor if the height of the step is increased from 50 mm to 200 mm.

5.22ps The beam's weight is $W = 200$ lb. Suppose that the attachment point of the wire is moved one foot to the left and its length BC adjusted so that the beam remains horizontal. The position of the support C remains fixed. Determine the reactions at A and the tension in the wire.

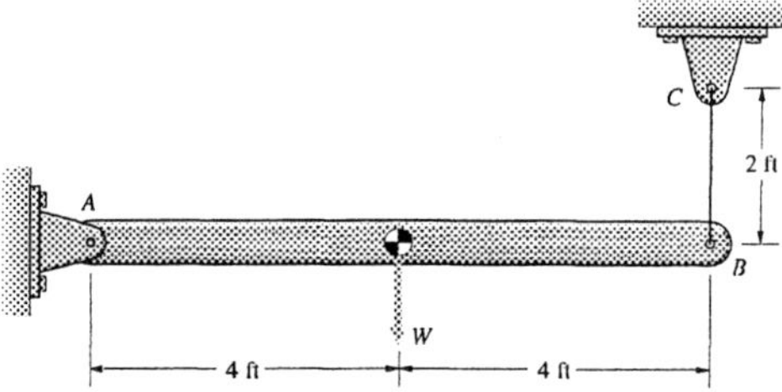

5.23ps The rectangular plate is held in equilibrium by the horizontal force F. The weight W acts at the midpoint of the plate. Assume that $b/h = 3$. Draw a graph of the angle α at which equilibrium occurs as a function of the ratio F/W for $0 < F/W < 5$.

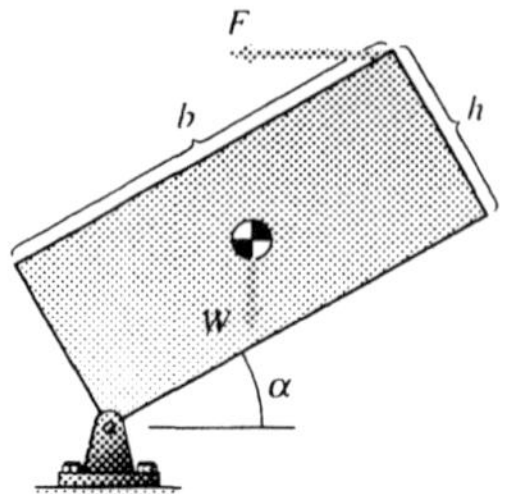

5.24ps The bar *AB* is of length *L*. Assume that its weight *W* acts at a distance $0.4L$ from point *A*. The length of the string is $0.5L$. What is the tension in the string?

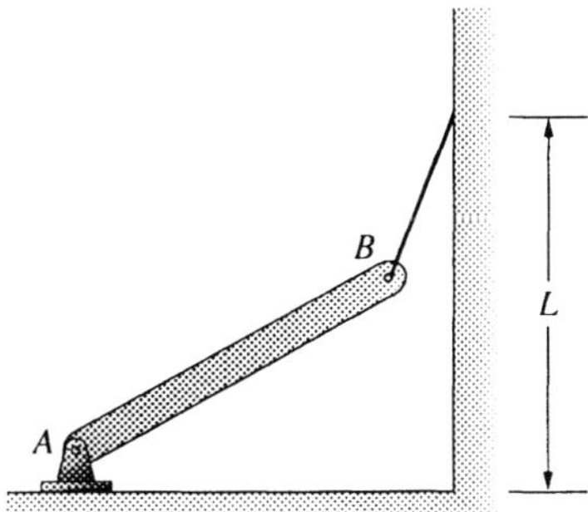

5.25ps The bar *AB* shown in Problem 5.24ps has a length *L* of 2 m and a weight *W* of 40 N. Assume that its weight *W* acts at a distance $0.4L$ from point *A*. The length of the string is $0.8L$. What is the tension in the string?

5.26ps The bar *AB* shown in Problem 5.24ps has a length *L* of 2 m and a weight *W* of 40 N. Assume that the weight *W* acts at a distance 0.6L from point *A*. Draw a graph of the tension in the string as a function of the length *d* of the string for $0.45L < d < 0.9L$.

5.27ps The mass of the truck is 4 Mg and all its wheels are locked. Assume that a counterweight of mass *m* is attached to the truck's front bumper 2.5 meters to the left of the truck's center of mass. Draw a graph of the maximum tension *T* that can be exerted by the truck's winch without lifting the front wheels off the ground as a function of the mass *m* for $0 < m < 400$ kg.

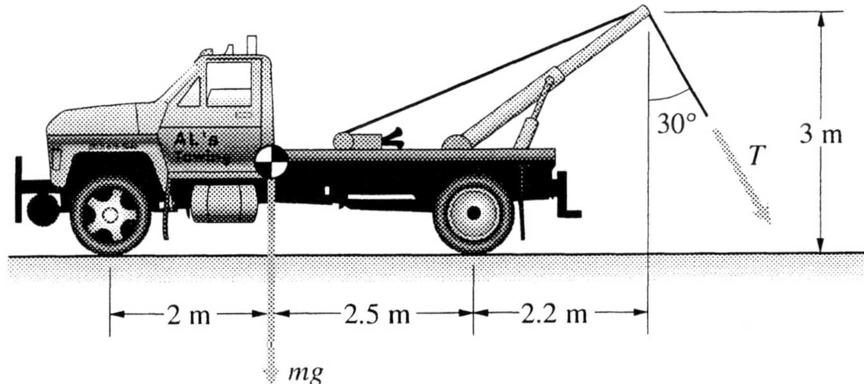

5.28ps Assume that in addition to the external loads shown, a 400-N downward force acts on the beam midway between the supports *A* and *B*.

(a) Draw the free-body diagram of the beam.
(b) Determine the reactions at A and B.

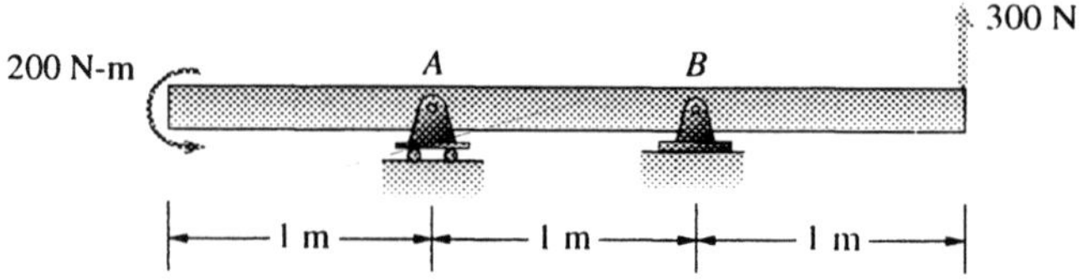

5.29ps Assume that in addition to supporting the 90-kg suspended object, a 1,000 N-m counterclockwise couple acts on the structure at A. Determine the reactions at A and B.

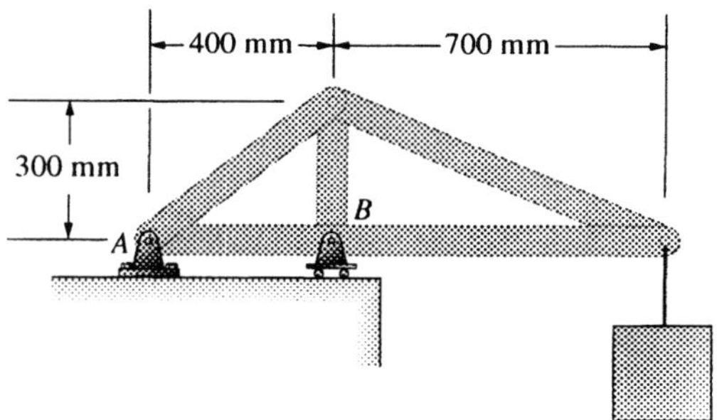

5.30ps Assume that the structure shown in Problem 5.29ps, in addition to supporting the 90-kg suspended object, is subjected to a counterclockwise couple of magnitude C at A.
(a) Determine the value $C = C_{\max}$ which causes the reaction B to be zero.
(b) Draw a graph of the reaction at B as a function of C for $0 < C < C_{\max}$.

5.31ps The mass of the car is 1500 kg and the dimensions $h = 0.4$ m and $b = 1.7$ m. The angle $\alpha = 15°$. The rear wheels are locked and the front wheels are free to roll. Determine A_x, A_y, and B.

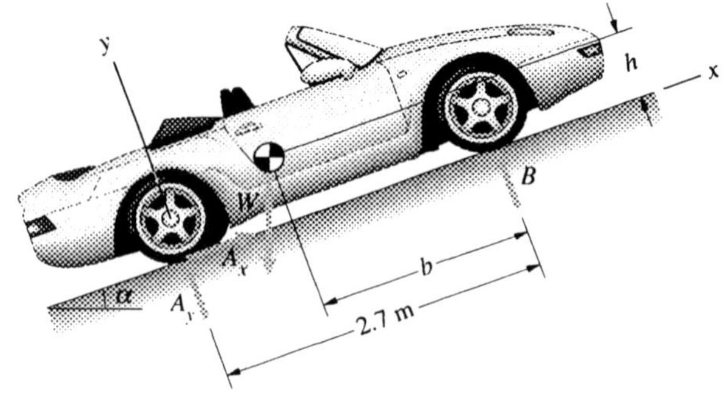

5.32ps The mass of the car shown in Problem 5.31ps is 1600 kg and the dimensions $h = 0.35$ m and $b = 1.8$ m. The front wheels are locked and the rear wheels are free to roll. Draw graphs of B_x, B_y, and A_y as functions of the angle α for $0 < \alpha < 20°$. From your plots, look at where the plots for B_x and B_y cross. What might be the importance of this point? Can B_x be greater than B_y in the real world?

5.33ps Assume that the angle of the slope the trailer is parked on is 3° instead of 15°. The owner of the trailer unhitches the trailer from the car and moves the car. She then lifts the hitch H back to the position shown. Determine the magnitude of the total force she must exert at H to prevent the trailer from moving: (a) if the trailer has no brakes; (b) if the trailer's wheels are locked.

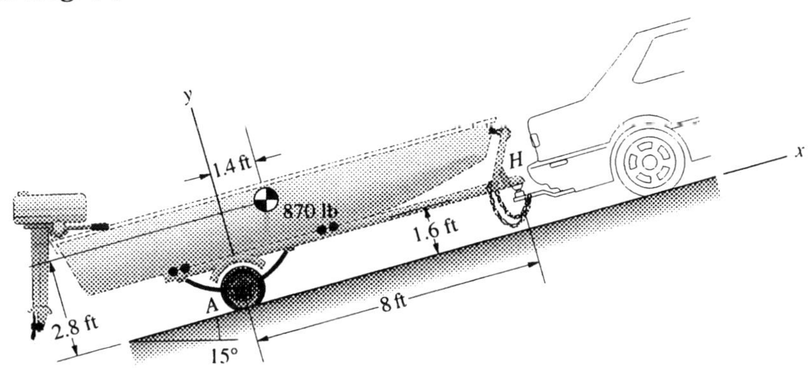

5.34ps The hydraulic actuator BC exerts a force at C that lies along the line from B to C. Treat A as a pin support. The mass of the suspended load is 3000 kg and the angle $\alpha = 30°$. Determine the force exerted by the actuator at C and the reactions at A.

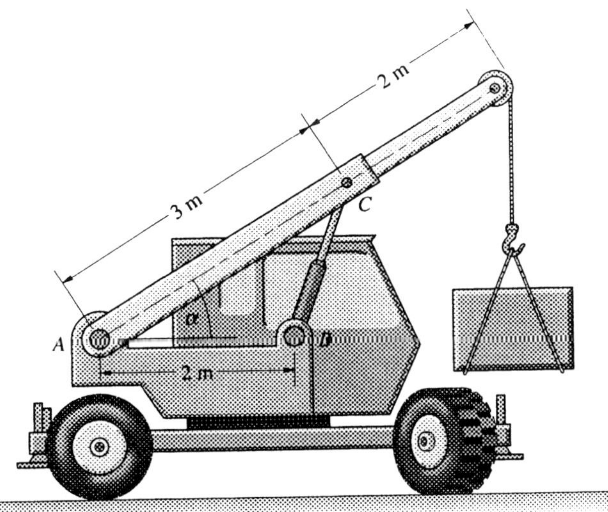

5.35ps In Problem 5.34ps, let the length of the extensible part of the boom (the part from the pin support C to the end of the boom) be denoted by L. Draw a graph of the force exerted by the actuator at C as a function of L for $0.5 < L < 3$ m.

5.36ps The horizontal rectangular plate is suspended by vertical cables. The plate's mass is 80 kg and its weight acts at its midpoint. Let the distance from the attachment point of cable C to the plate's lower right-hand corner be denoted by d. Draw graphs of the tensions in the cables A, B, and C as functions of the distance d for $0.1 < d < 1.9$ m.

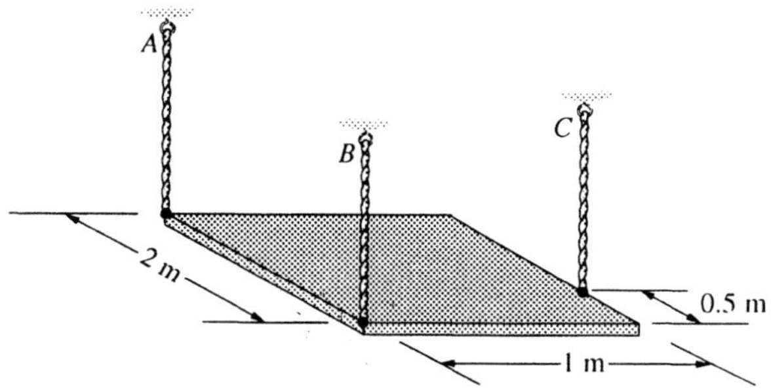

5.37ps Consider the suspended plate described in Problem 5.36ps. Assume that cables A and C are positioned as shown. Determine the range of positions along the plate's left edge at which the vertical cable B can be attached and have the plate remain in equilibrium in the horizontal position shown.

5.38ps The 80-lb bar is supported by a ball and socket at A, the smooth wall it leans on at B, and the cable BC. The bar's weight acts at the point 3 ft along the bar from point A. Determine the tension in cable BC and the reactions at A.

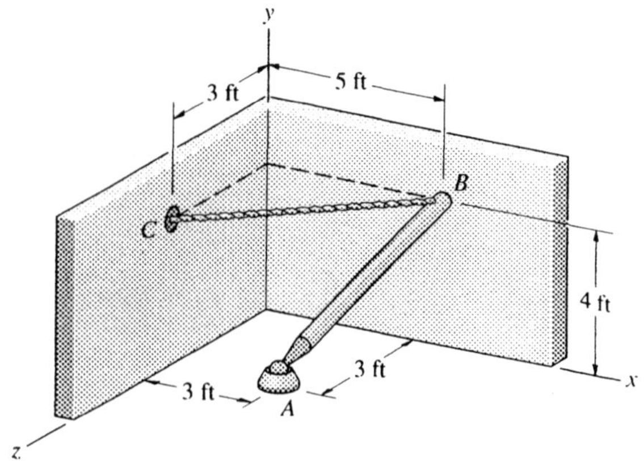

Chapter 6: Structure in Equilibrium

6.1ps Assume that in addition to the 2-kip external force acting on the truss, a 1-kip horizontal force to the right acts at point *C*. Determine the axial forces in the members of the truss.

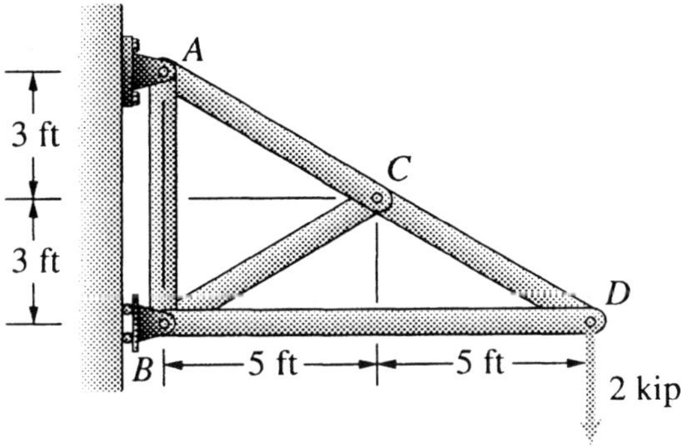

6.2ps In the Pratt truss shown, assume that the forces *F* acting at the joints *B*, *C*, and *D* are changed to $2F$, $3F$, and $3F$, respectively.
(a) Draw the free-body diagram of the entire truss and determine the reactions and *A* and *E*.
(b) Determine the axial forces in the members of the truss.

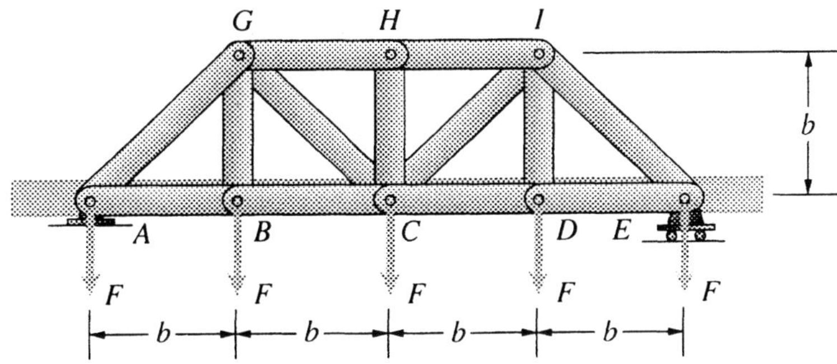

6.3ps Suppose that the lengths of members *AB* and *AC* are changed so that the 30° angle increases to 45° and the 20° angle is decreases to 5°. The weight $W = 500$ lb.
(a) Draw the free-body diagram of joint *A*.

(b) Determine the axial forces in members *AB* and *AC*.

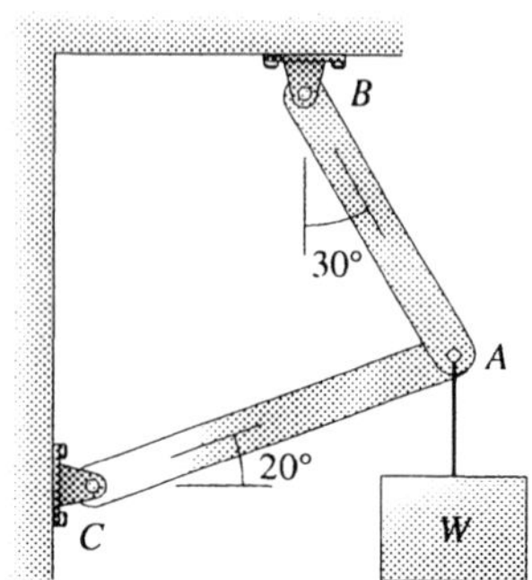

6.4ps For the truss shown in Problem 6.3ps, let $W = 1{,}000$ lb.
(a) Determine the axial forces in members *AB* and *AC* with the geometry as shown.
(b) If the lengths of members *AB* and *AC* are varied so that the 20° angle remains constant and the 30° angle increases to 45°, what are the axial forces in members *AB* and *AC*?
(c) If the lengths of members *AB* and *AC* are varied so that the 20° angle remains constant and the 30° angle increases to 60°, what are the axial forces in members *AB* and *AC*?

6.5ps In addition to the forces $F_1 = 60$ N and $F_2 = 40$ N shown, suppose that a 30-N horizontal force pointing toward the left acts at *B*.
(a) Draw the free-body diagram of the entire truss and determine the reactions at *B* and *C*.
(b) Determine the axial forces in the members of the truss. Indicate whether they are in compression (C) or tension (T).

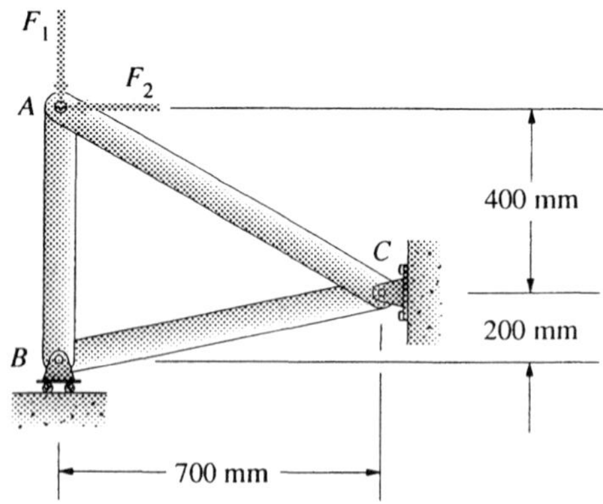

6.6ps Consider the truss shown in Problem 6.5ps. Suppose that in addition to the forces $F_1 = 60$ N and $F_2 = 40$ N, a horizontal force *F* pointing toward the left acts at *B*.

(a) Determine the maximum value of F for which the roller at B remains in contact with the surface on which it rests.
(b) If F has the value determined in part (a), what are the axial forces in the members of the truss? Indicate whether they are in compression (C) or tension (T).

6.7ps Consider the truss shown in Problem 6.5ps. Suppose that $F_1 = 250$ lb and the direction of the force F_2 is reversed.
(a) Determine the maximum value of F_2 for which the roller at B remains in contact with the surface on which it rests.
(b) If F_2 has the value determined in part (a), what are the axial forces in the members of the truss? Indicate whether they are in compression (C) or tension (T).

6.8ps Assume that in addition to the force $F = 600$ lb, a horizontal 300-lb force pointing toward the left is applied at B. Determine the axial forces in the members of the truss. Indicate whether they are in compression (C) or tension (T).

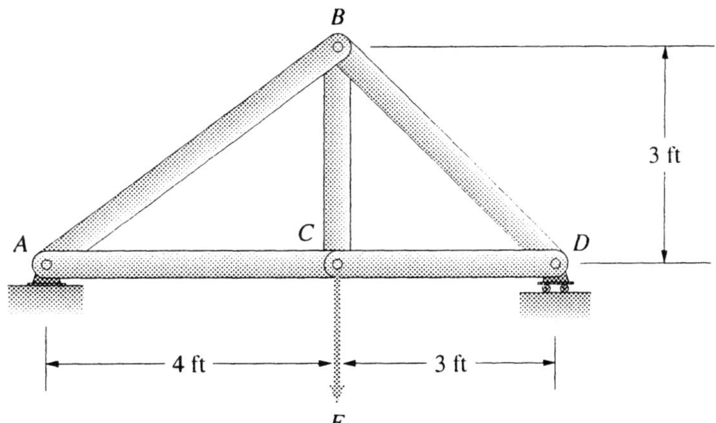

6.9ps Consider the truss shown in Problem 6.8ps. Assume that in addition to the force $F = 600$ lb, a horizontal force G pointing toward the left is applied at B. Determine the maximum value $G = G_{max}$ for which the roller at D remains in contact with the surface on which it rests. Draw a graph of the axial force in member AB as a function of G for $0 < G < G_{max}$.

6.10ps Consider the truss shown in Problem 6.8ps. The horizontal 300-lb force is still acting at B. Assume that the triangle ABD remains as shown in the figure, but that the location of the joint C along the line from A to D is allowed to vary. The distance from A to C can vary over the range (1 ft $< AC <$ 6 ft). The force $F = 600$ lb remains vertical. What distance AC minimizes the axial load in member BC?

6.11ps Assume that in addition to the external force $F = 12$ kN shown, a horizontal force $G = 20$ kN pointing toward the right acts at B. Determine the axial forces in the members of the truss. Indicate whether they are in compression (C) or tension (T).

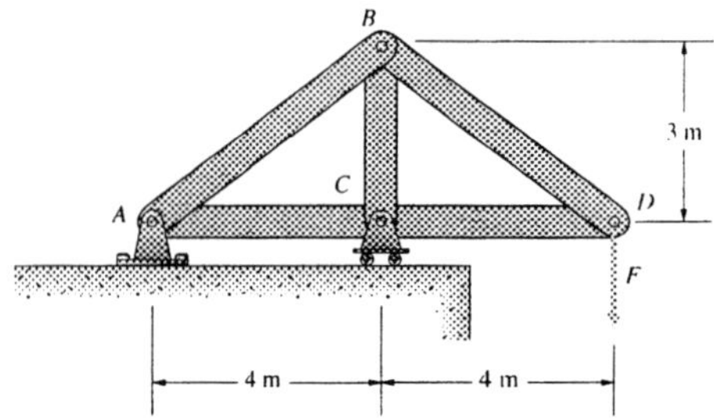

6.12ps Each member of the truss shown in Problem 6.11ps will safely support a tensile force of 150 kN and a compressive force of 30 kN. Can a horizontal force applied at *B* increase the maximum downward force *F* that the truss can safely support? If so, determine the horizontal force *G* that maximizes the force *F* that the truss can safely support. Determine the axial forces in the members of the truss for this loading.

6.13ps Suppose that the truss shown in Problem 6.11ps is modified so that the length of the vertical member *BC* is 4 m. If the force $F = 10$ kN, and there are no other loads acting on the truss, what are the axial forces in the members of the truss?

6.14ps For the truss shown in Problem 6.11ps, suppose that the length of the vertical member BC is allowed to vary over the range (1 m < BC < 8 m). The force $F = 10$ kN and there are no other loads acting on the truss. Draw a graph of the axial force in member *AB* as a function of the length of member BC. Determine the length for member BC within the allowable range which minimizes the axial force in member *AB*.

6.15ps Suppose that the vertical forces applied at joints *G*, *H*, and *I* of the two trusses are *F*, 2*F*, and 3*F*, respectively.
(a) In which truss does the largest tensile force occur? Determine its value and the member in which it occurs.
(b) In which truss does the largest compressive force occur? Determine its value and the member in which it occurs.

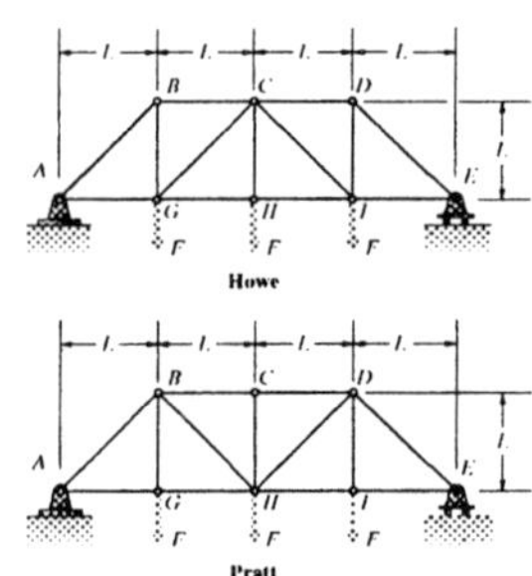

6.16ps The sum of the two forces F_1 and F_2 is 20 kip to the right. Find the combination of values for F_1 and F_2 which minimizes *the* load in member *BG*. Strategy: Choose values for F_1.

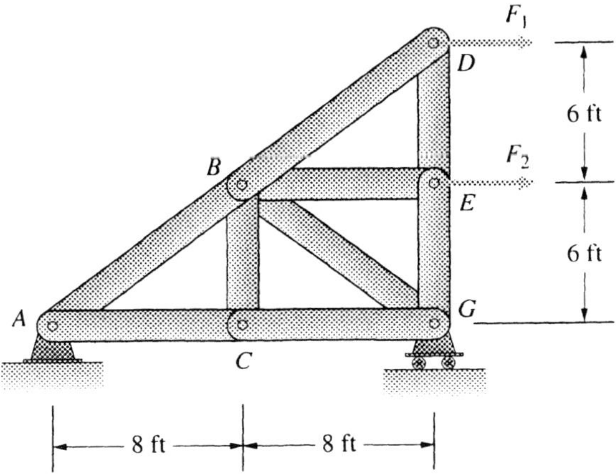

6.17ps The sum of the two forces F_1 and F_2 acting on the truss shown in Problem 6.16ps is 20 kip to the right. Find the combination of values for F_1 and F_2 which minimizes the load in member *AB*.

6.18ps Let $F = 3$ kN. Is it possible to decrease the load in the most heavily loaded member by applying a vertical load (up or down) at *B*? If so, determine the additional load at *B* which causes the load in the most heavily loaded member to be smallest.

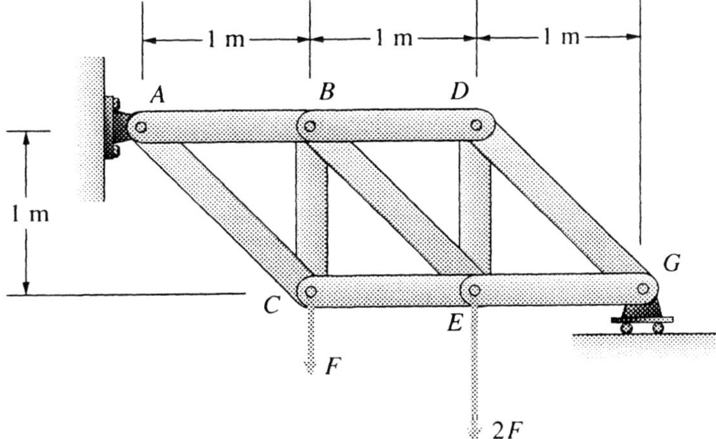

6.19ps The stretcher exerts equal weights $W/2$ on the truss at *A* and *B*. It is proposed that the truss be redesigned by replacing members *FG*, *DF*, and *CF* by a single member *CG*. Calculate the axial forces in the members of the original truss and in the members of the redesigned truss and compare them. (Your answers will be in terms of W.)

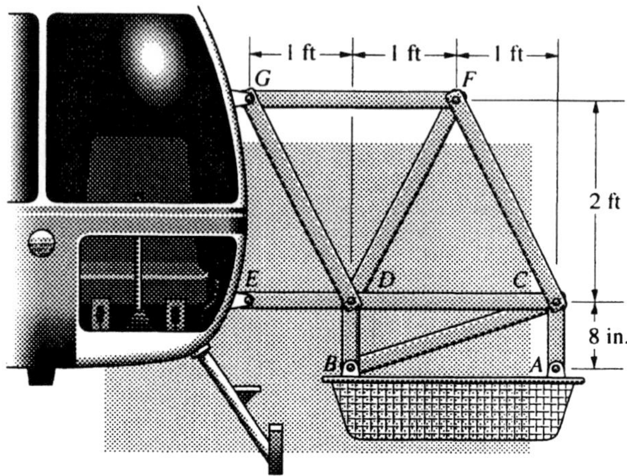

6.20ps Assume that in addition to the weight W shown, a second weight W is suspended from joint D. Determine the axial forces in the members of the truss. (Your answers will be in terms of W.)

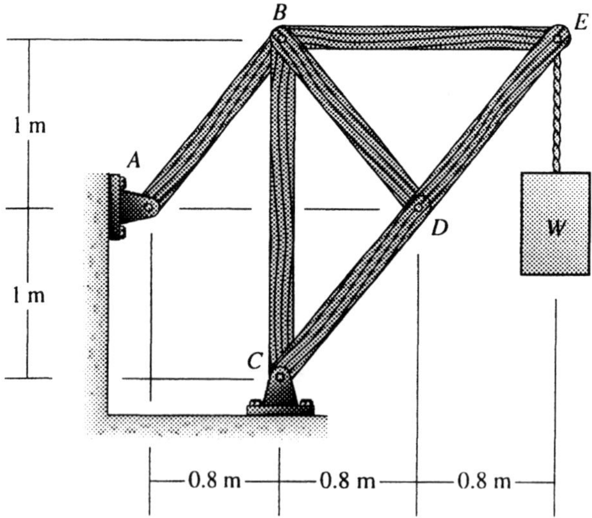

6.21ps Assume that in addition to the forces $F_1 = 400$ lb and $F_2 = 550$ lb, a horizontal force $F_3 = 300$ lb pointing toward the left is applied at joint A. Determine the axial forces in members AB, AC, AE, and BC.

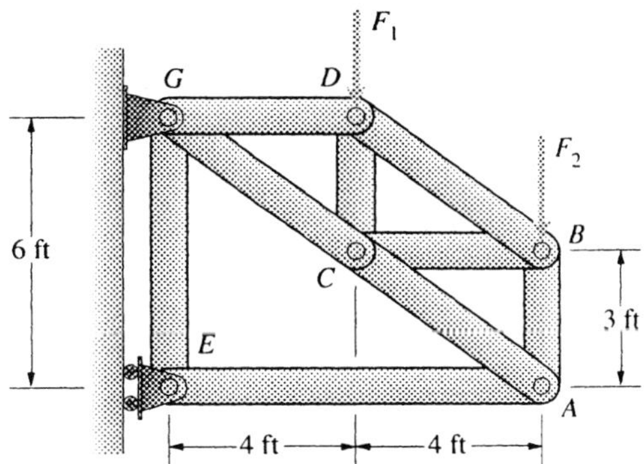

6.22ps Each member of the truss will safely support a tensile force of 4 kN and a compressive force of 2 kN. Based on this criterion, what is the largest mass m that can safely be supported?
Strategy: Assume a 1-kN load and scale the results.

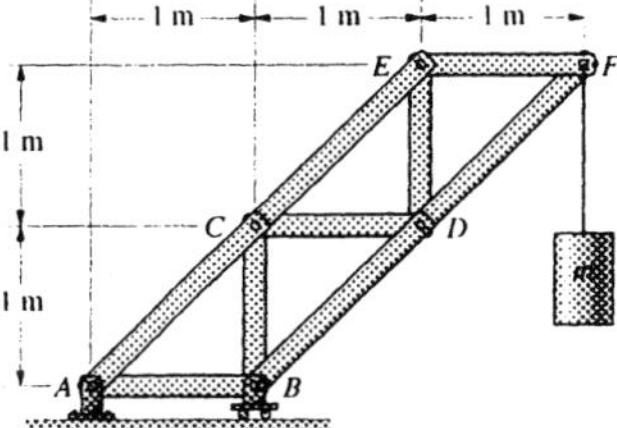

6.23ps Each member of the truss shown in Problem 6.22ps will safely support a tensile force of 4 kN and a compressive force of 2 kN. Would a horizontal force G applied either at C or at E increase the largest mass m that can safely be supported? If so, find the value and point of application for G that would allow the maximum mass to safely be supported.

6.24ps Design a nine-member truss with no zero-force members that is attached at A and B and supports the forces at C and D. Determine the axial force in each member.

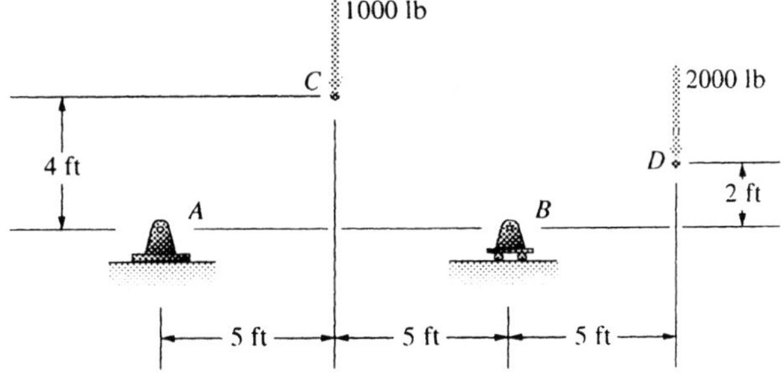

6.25ps Assume that, in addition to the two forces at C and D, there is a 3-kN force that is horizontal, points to the right, and acts at a point E. Point E is located at the intersection of a horizontal line drawn from A with a vertical line drawn from D. Design a seven-member truss with no zero-force members to support the three forces and calculate the axial force in each member.

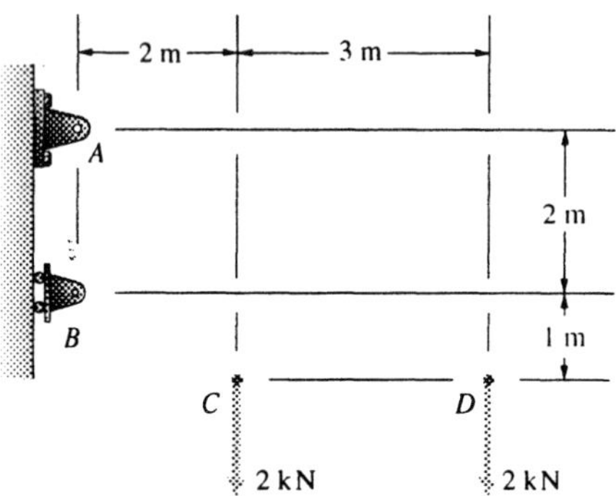

6.26ps Assume that instead of the 100-kN force, four 25-kN downward forces act at joints J, K, L, and M. Determine the axial force in member CJ and compare your result to the result obtained in Example 6.4 (page 291).

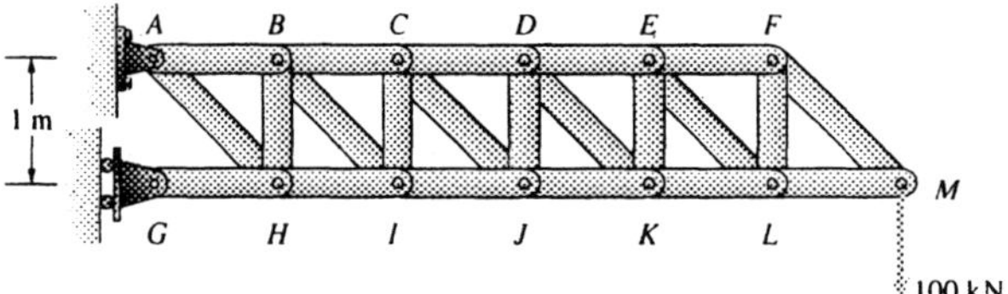

6.27ps Assume that in addition to the loading shown, a downward force $2F$ acts at D. Determine the axial forces in members DG, BE, and CG.

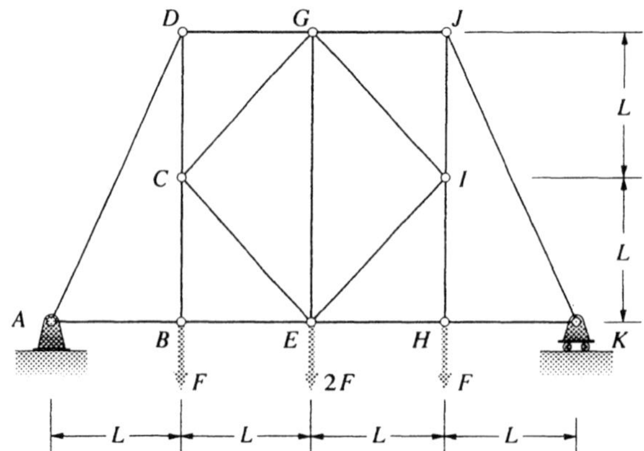

6.28ps Assume that in addition to the loading shown, a 150-kN downward force acts at *G*.
(a) Draw the free-body diagram of the entire truss and calculate the reactions at *B* and *F*.
(b) Use the method of sections to determine the axial forces in members *CE*, *BE*, *BD*, *GF*, *GH*, and *HF*.

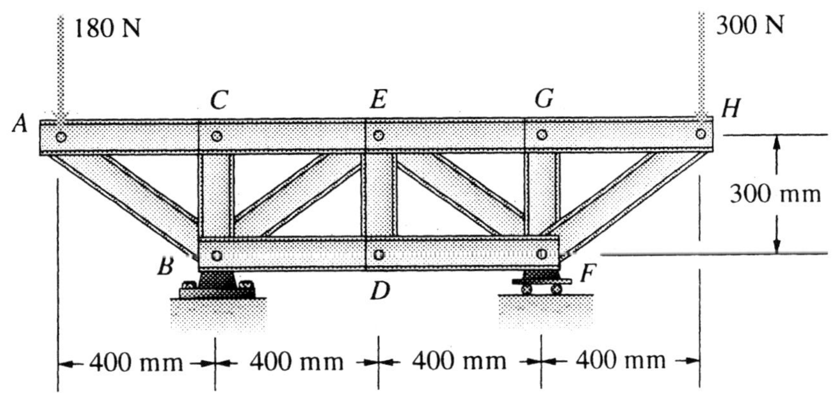

6.29ps Assume that in addition to the loading shown, a 5-kip downward force acts at *C*.
(a) Draw the free-body diagram of the entire truss and determine the reactions at *A* and *C*.
(b) Use the method of sections to determine the loads in members *CF* and *CH*.

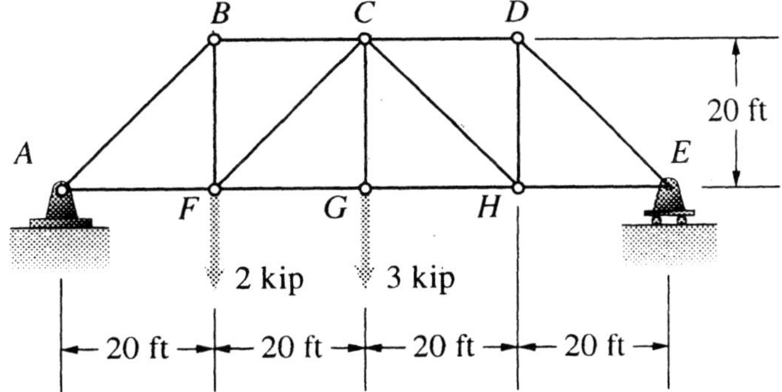

6.30ps Assume that in addition to the loading shown in the figure of the truss in Problem 6.29ps, a 5-kip downward force acts at *C*. Use the method of joints to determine the axial force in member *CG*.

6.31ps Assume that the loading on the Pratt bridge truss is modified by removing the load at J and increasing the load at K to $2F$. Use the method of sections to determine the axial force in member JK.

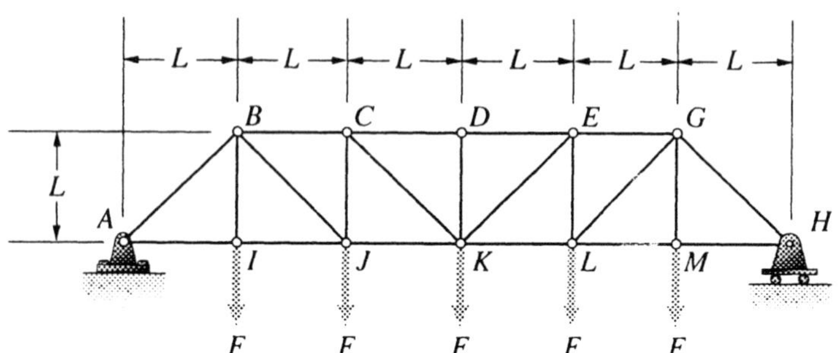

6.32ps Suppose that in addition to the loading shown on the figure of the Pratt bridge truss in Problem 6.31ps, you can apply a horizontal force F_H at H. The load at J is zero and the load at K is $2F$. Find the value of F_H that minimizes the axial force in member JK.

6.33ps Assume that the downward forces on the roof truss are changed to the following values: At B, 4 kN; at C, 3 kN; at D, 1 kN; at E, 3 kN, and at F, 4 kN.
(a) Draw the free-body diagram of the roof truss with its new loading.
(b) Use the method of joints to determine the axial forces in members CJ and BI.
(c) Use the method of sections to verify the results of part (b).

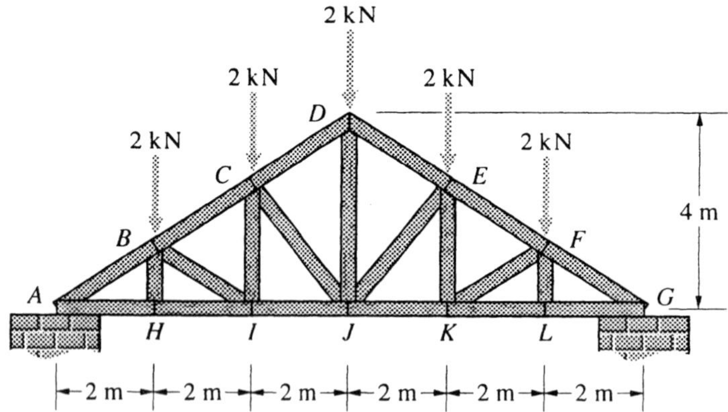

6.34ps For the roof truss shown in Problem 6.33ps, treat A as a pin support and G as a roller support. Assume that in addition to the vertical forces shown in the original figure, 1-kN horizontal wind forces directed toward the right act at B, C, and D.
(a) Draw the free-body diagram of the truss with its new loading.
(b) What is the axial force in member CJ with and without the wind loading?

6.35ps Assume that in addition to the loading shown, the truss is subjected to a

15-kip downward force at *E*. Determine the changes in the axial forces in members *EF*, *EG*, *DE*, and *CE* caused by the addition of the vertical force.

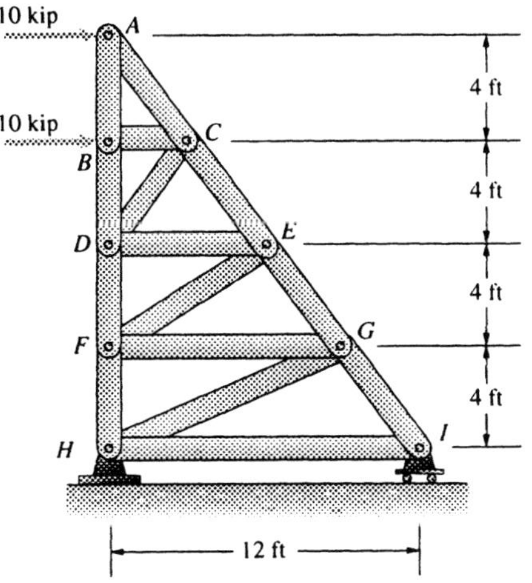

6.36ps You are to suggest a modified loading on the truss shown in Problem 6.35ps using the following guidelines: (1) The total force on the truss must be 20 kip to the right. (2) Loads may be applied only at joints *A*, *B*, *D*, and/or *F*. Your objective is to minimize the axial force in member *GI*.

6.37ps Assume that the truss is redesigned by adding a new member *GL* and removing member *JK*. Compare the axial load in member JK of the original truss with the load in member *GL* of the redesigned one.

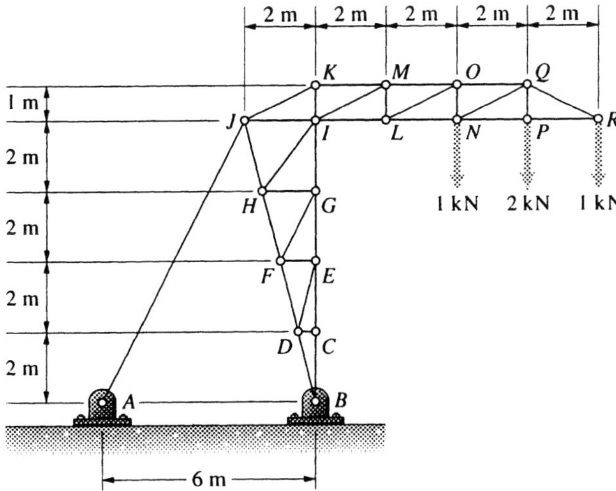

6.38ps Compare the axial forces in members *HI*, *GI*, and *KI* of the original truss described in Problem 6.37ps with the forces in the same members of the redesigned truss.

6.39ps The weight $W = 600$ lb. Assume that the position of the attachment D remains fixed and the length L of the cable from A to D is varied. Draw a graph of the tension in cable AD as a function of L for $1 < L < 8$ ft.

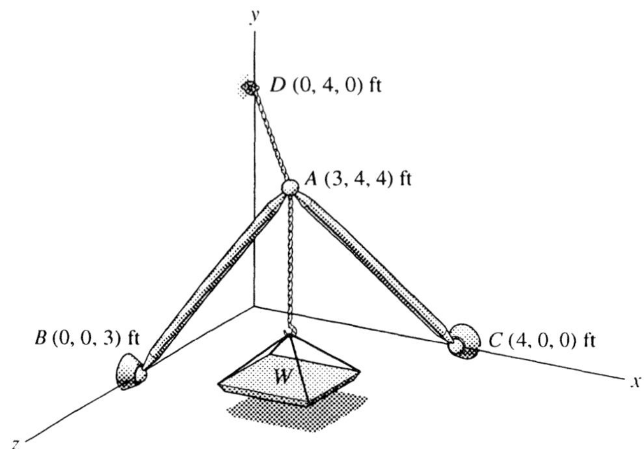

6.40ps For the system shown in Problem 6.39ps, the weight $W = 600$ lb. Assume that the position of the attachment D remains fixed and the length L of the cable from A to D is varied over the range $1 < L < 7$ ft.
(a) Determine the axial force in members AB and AC, and find the tension in AD.

6.41ps The space truss is being considered as a support for a large (200 lb) studio television camera. It is required that triangle BCD be equilateral and that the lengths $AB = AC = AD$. Assume that $BC = 2$ ft and 2ft. Draw graphs of the axial forces in members AB and AC as functions of the length of member AB.

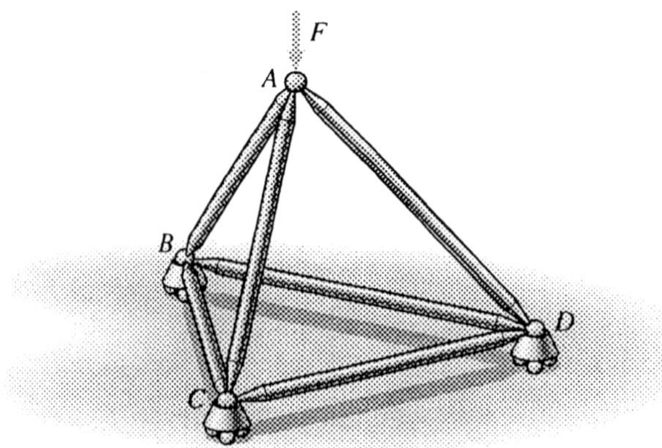

6.42ps The weight $W = 100$ lb. Assume that you can vary the location of point B, changing the length of member BE so that member CEG remains horizontal and the location of point E remains fixed. Draw a graph of the axial force in member BE as a function of the distance BC for $3 < BC < 8$ in.

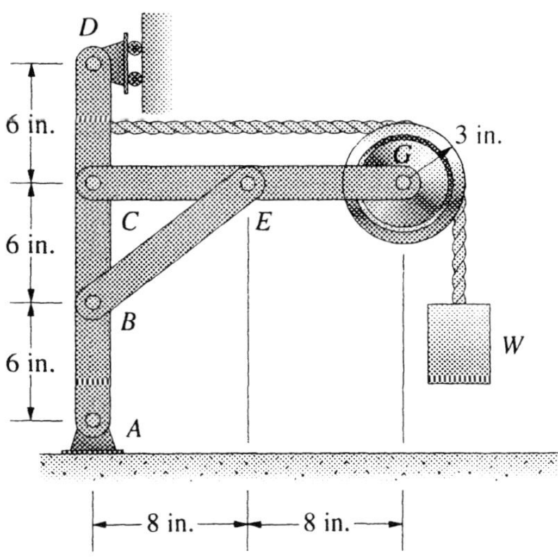

6.43ps The pliers shown have a mechanical advantage of 10.1 (See Example 6.10, page 314). If you vary the length of the link AB by keeping point A fixed and moving point B parallel to the line BD, what location for point B relative to its original position increases the mechanical advantage of the pliers to 11?

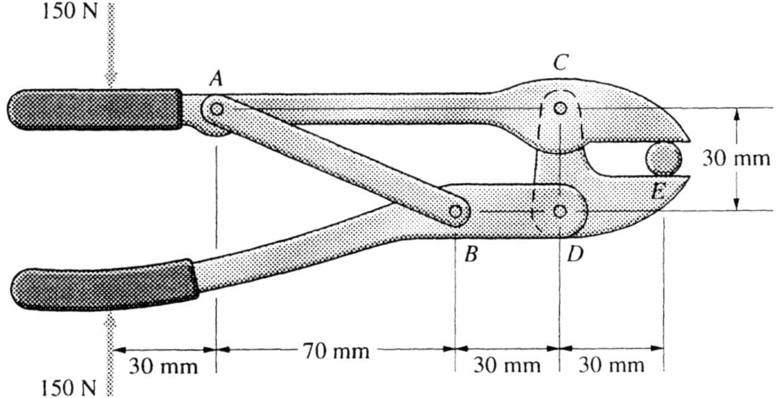

6.44ps The frame ABC is subjected to a force and a couple.
(a) Determine the forces on member BC.
(b) Suppose that you increase the 100 N-m couple to the maximum value it can have without causing the support at C to lift off the surface on which it rests. What are the resulting reactions at the built-in support A?

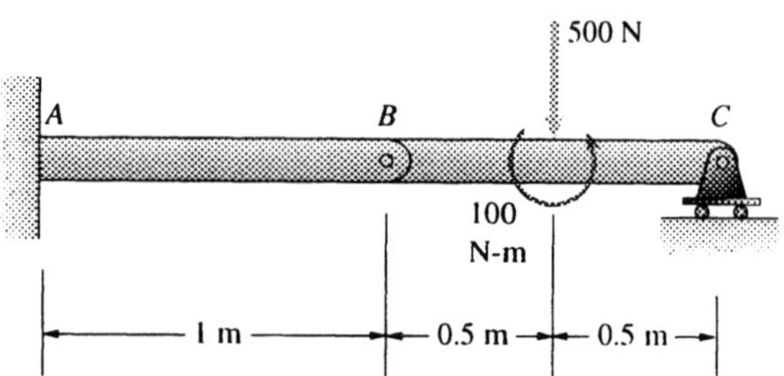

6.45ps The largest lifting force that the jack safely can exert at *C* is limited by the pin support at *A*, which will support a force of 20-kN magnitude. What is the magnitude of the largest lifting force, and what is the resulting axial force in the hydraulic actuator *BD*?

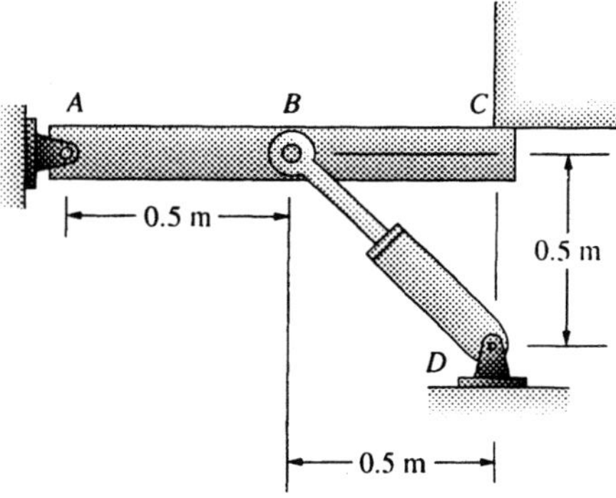

6.46ps Assume that in addition to the 800-N force shown, a horizontal 200-N force to the right acts at the midpoint of member *AB*. Determine the forces acting on the members of the frame.

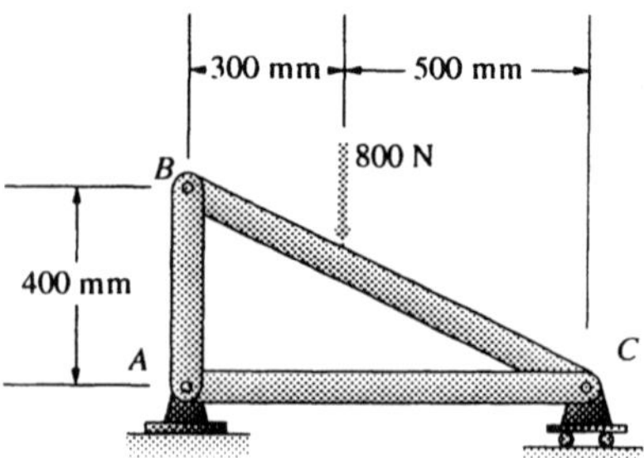

6.47ps Suppose that you want to redesign the frame by lengthening member *CE*, keeping point *E* fixed and moving point *C* to a new location somewhere between point B and a point 0.5 meters below point D. Let H be the height of point C above point B. Draw graphs of the axial force in member CE and the magnitude of the reaction at the pin support D as functions of H for $0 < H < 1.5$ m.

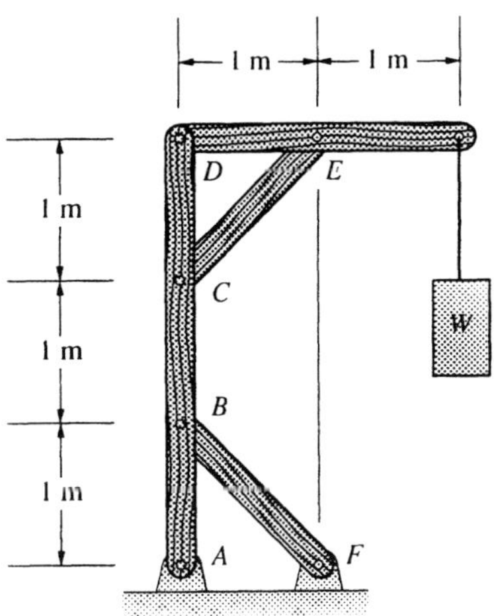

6.48ps Suppose that the 18-lb downward force at B is changed to a 64-lb upward force. What are the reactions at A and C?

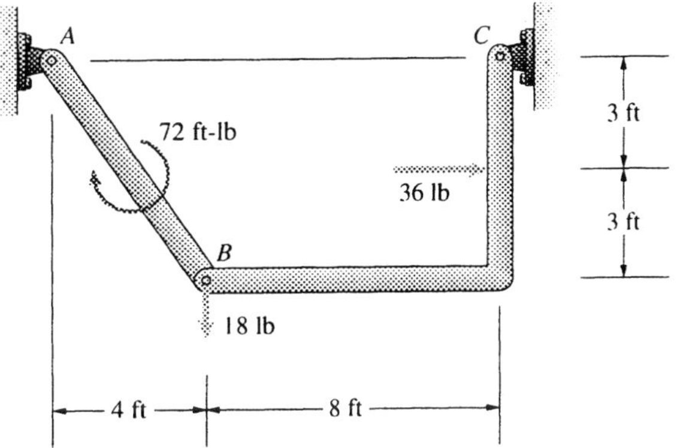

Chapter 7: Centroids and Centers of Mass

7.1ps What are the x and y coordinates of the centroid of the area shown if the curve $y = x^2$ is replaced by the curve $y = x^3$?
338.)

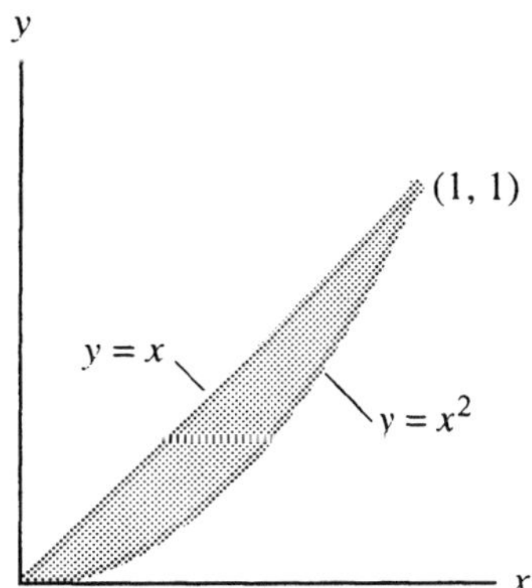

7.2ps In the figure of Problem 7.1ps, replace the curve $y = x^2$ by the curve $y = x^n$. Determine the x and y coordinates of the centroid of the area. Show that the x coordinate of the centroid approaches $\tilde{x} = 2/3$ and the y-coordinate of the centroid approaches $\tilde{y} = 1/3$ as n. Discuss why this should be expected. Strategy: Examine the succession of areas as $n \to \infty$ increases.

7.3ps What is the x coordinate of the centroid of the line L shown if the curve $y = x^2$ is replaced by the curve $y = x^5$?

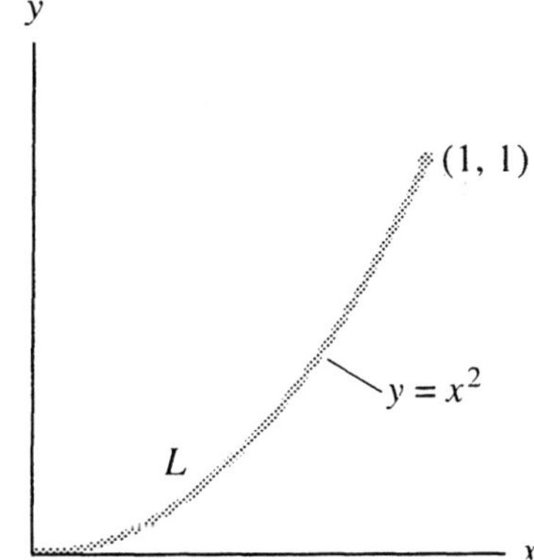

7.4ps In the figure of Problem 7.3ps, replace the curve $y = x^2$ by the curve $y = x^n$. What is the approximate value of the x coordinate of the centroid of the line for large positive values of n?
Strategy: Examine the succession of curves as n increases.

7.5ps What is the x coordinate of the centroid of the area shown if the curve $y = x^2$ is replaced by the curve $y = x^{1/2}$?

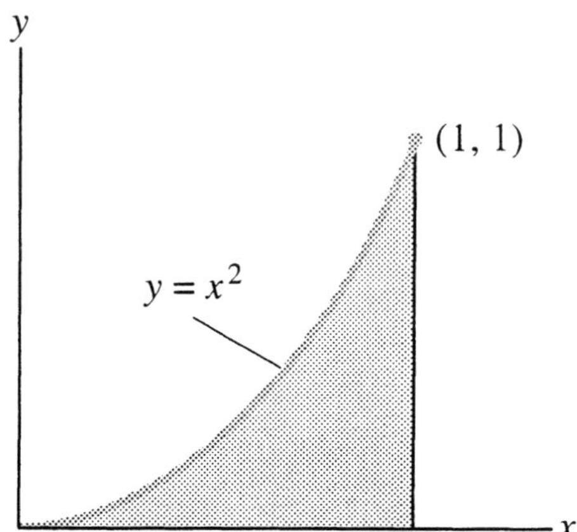

7.6ps What are the x and y coordinates of the centroid of the area of the part of the board to the left of the curve $y = x + x^3$?

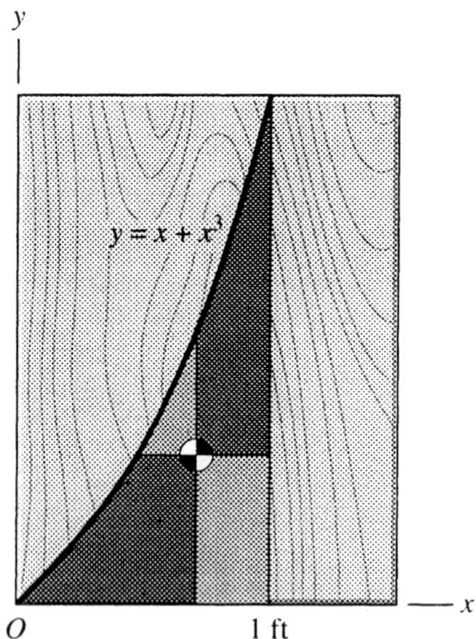

7.7ps A volume is obtained by revolving the curve $y = x^{1/2}$, $0 \le x \le 1$, about the x axis. A second volume is obtained by revolving the curve $y = (1/4)x^2$, $0 \le x \le 1$, about the x axis. Determine the centroid of the volume obtained by subtracting the second volume of revolution from the first.

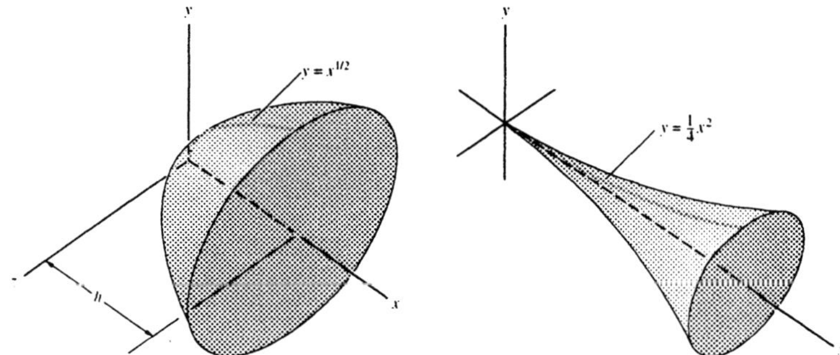

7.8ps Suppose that the area shown has a circular hole of radius $R/2$ centered at $(b + c/2,\ R)$. Draw the new area to scale and determine the x and y coordinates of its centroid.

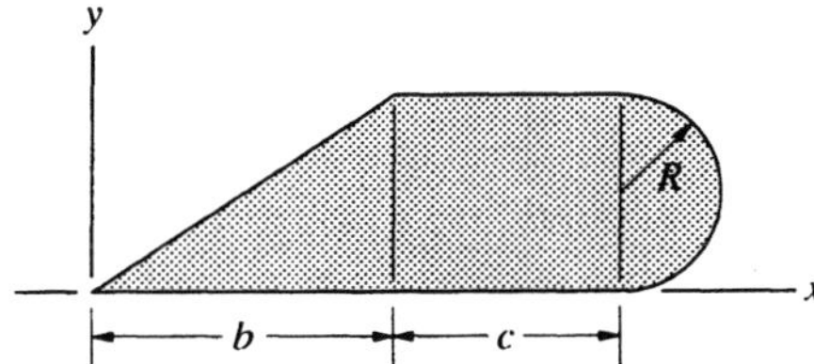

7.9ps Suppose that the volume shown has a hemispherical cap of radius R placed on its flat cylindrical end. Determine the centroid of the new composite volume.

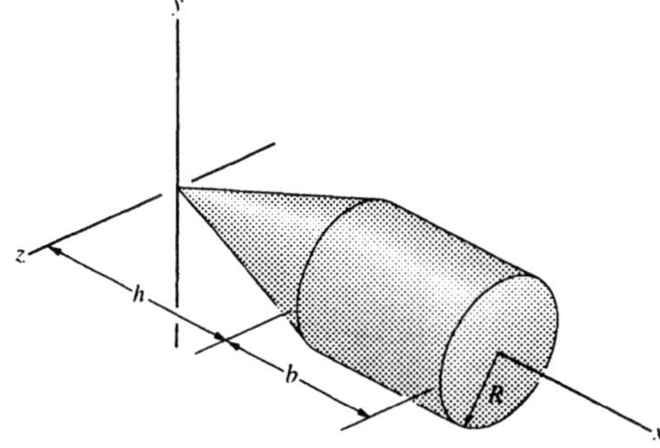

7.10ps The L-shaped part is composed of two homogeneous bars. Bar 1 is made of balsa wood with a mass density of 200 kgm^3. Bar 2 is made of gold with a mass density of 19,300 kg/m^3. Determine the center of mass of the L-shaped part. What is the distance from the center of mass of the L-shaped part to the center of mass of the gold bar?

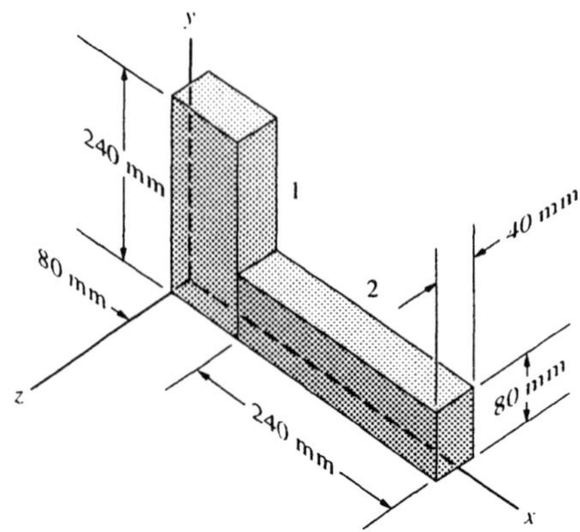

7.11ps Suppose that you want to position the centroid of the homogeneous plate at (10, 9, 2) in. by cutting a hole five inches in diameter in the plate. Is this possible? If so, where would you center the 5-in. hole?

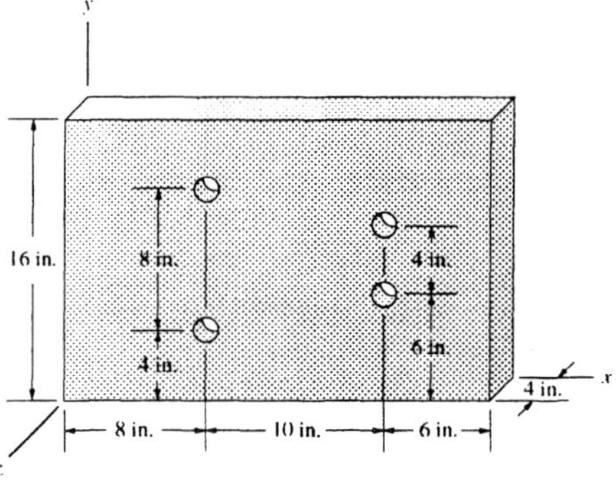

7.12ps The cylindrical tube is made of aluminum with a mass density of 2,700 kg/m^3. The cylindrical plug is made of gold with a mass density of 19,300 kg/m^3. Determine the coordinates of the center of mass of the composite object.

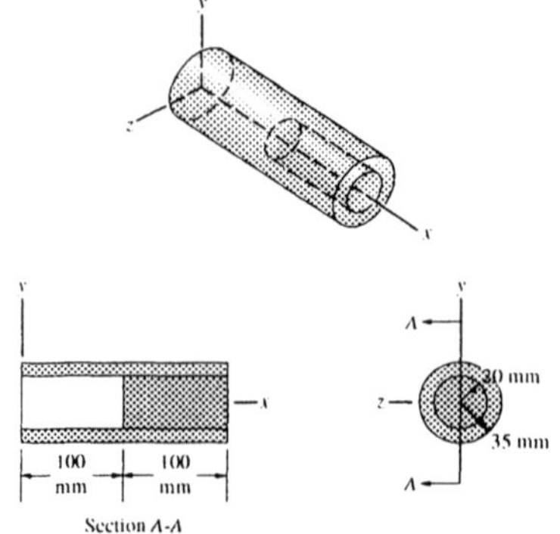

7.13ps Suppose that a second hole with a 100-mm radius is cut in the homogeneous plate. The new hole is centered 150 mm to the right of *A* and 200 mm above *A*. Draw the new plate to scale and determine the position of its center of mass. If the mass of the plate is 40 kg, what are the reactions at *A* and *B*?

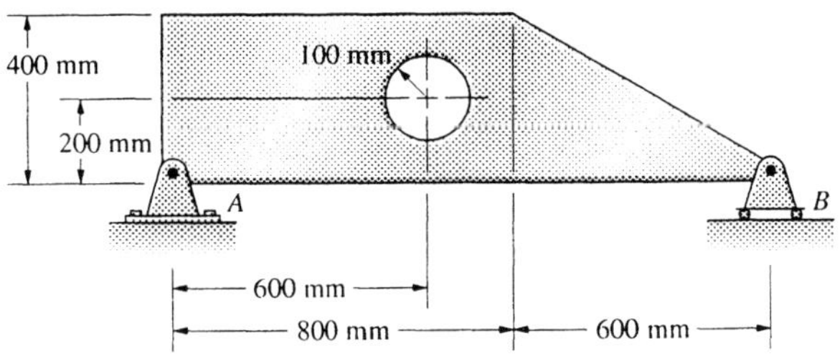

7.14ps Suppose that you remove the part of the plate shown to the right of the vertical line $x = 200$ mm. What are the coordinates of the center of mass of the remaining part?

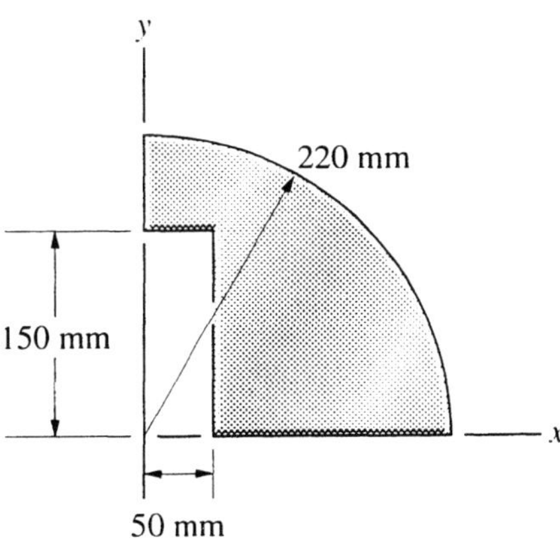

7.15ps Suppose that the homogeneous steel plate shown is modified by removing the part below and to the right of the straight line $y = x - 60$ mm. Draw the modified part to scale and determine the coordinates of its center of mass.

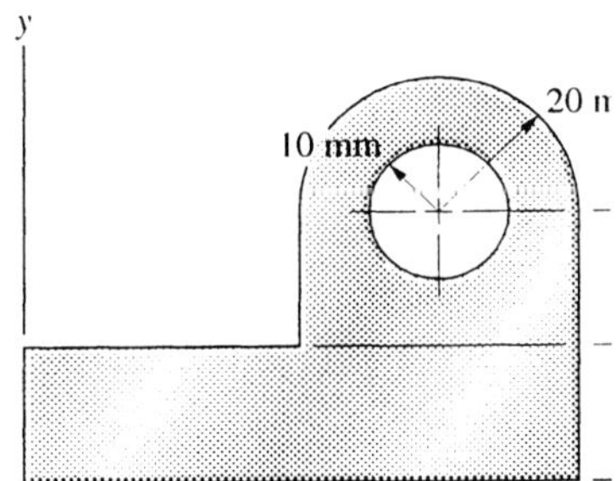

Chapter 8: Moments of Inertia

8.1ps For a coordinate system x'-y' parallel to the x-y system and with its origin at the centroid of the triangle, use integration to determine the moments of inertia $I_{x'}$ and $I_{y'}$.

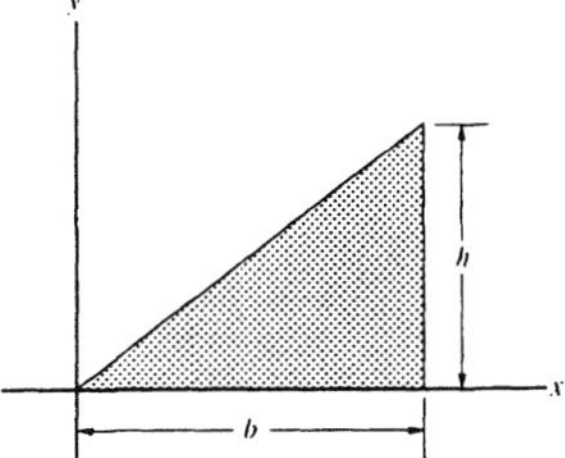

8.2ps For a coordinate system x'-y' parallel to the x-y system and with its origin at the centroid of the composite area, use the parallel-axis theorems to determine the moments of inertia $I_{x'}$ and $I_{y'}$. (See Example 8.4, page 398.)

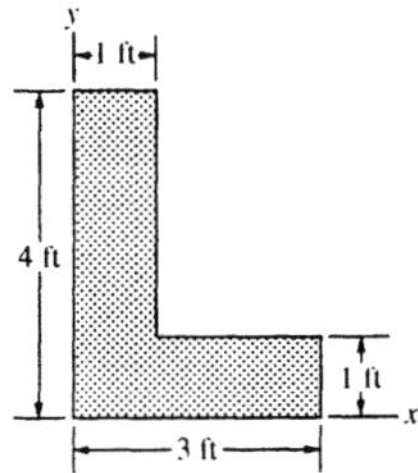

8.3ps For a coordinate system x'-y' parallel to the x-y system and with its origin at the centroid of the composite area, use the parallel-axis theorems to determine $I_{x'}$, $I_{y'}$, and $I_{x'y'}$.

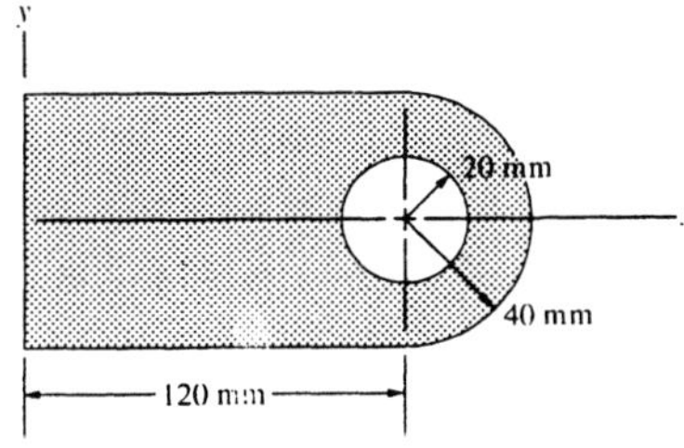

8.4ps Assume that the area shown in Problem 8.3ps has a second hole of 20-mm radius centered at (30, 0) mm. Draw the new plate to scale and determine I_y and k_y.

8.5ps Consider the composite area shown in Problem 8.4ps. For a coordinate system x'-y' parallel to the x-y system and with its origin at the centroid of the area, determine the moments of inertia $I_{x'}$ and $I_{y'}$.

8.6ps Consider the composite area shown in Problem 8.4ps. For a coordinate system x'-y' parallel to the x-y system and with its origin at the centroid of the area, determine the product of inertia $I_{x'y'}$.

8.7ps Two homogeneous slender bars, each of length L and mass m, are welded together as shown to form an L-shaped object. The axis L_o is perpendicular to the two bars. Determine the mass moment of inertia of the object about the axis that is parallel to L_o and passes through the center of mass of the object, and (b) parallel to L_o and passes through the center of mass of the lower bar.

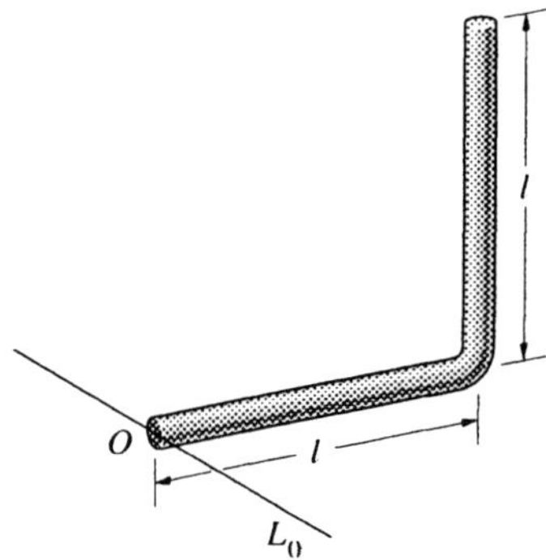

8.8ps Consider the L-shaped object described in Problem 8.7ps. Determine the mass moment of inertia of the object about the axis that is parallel to L_0 and passes through the point where the two bars are welded together.

8.9ps The bar shown is of mass m and length l. Suppose that a second identical bar rotated at a clockwise angle θ relative to the line L is welded to the bar shown so that the two bars form an X. (They are welded at their common center of mass.) The mass moment of inertia of the X about the axis perpendicular to the bars and through the point where they are welded is of the form $I = cmL^2$. Draw a graph of the valus of c as a function of θ for $5° \leq \theta \leq 45°$.

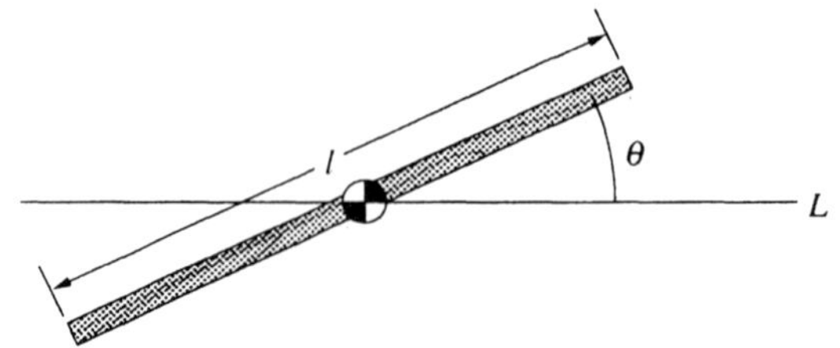

8.10ps You are evaluating different shapes with equal areas as candidates for the cross-sectional area of a beam. You decide to consider a hollow cylinder with outer radius $R_o = 40$ mm and inner radius $R_i = 20$ mm as a candidate. Compare the moment of inertia I_x (x goes through the centroid in both cases) of the hollow cylindrical cross-section to that of a solid square cross-section having the same area.

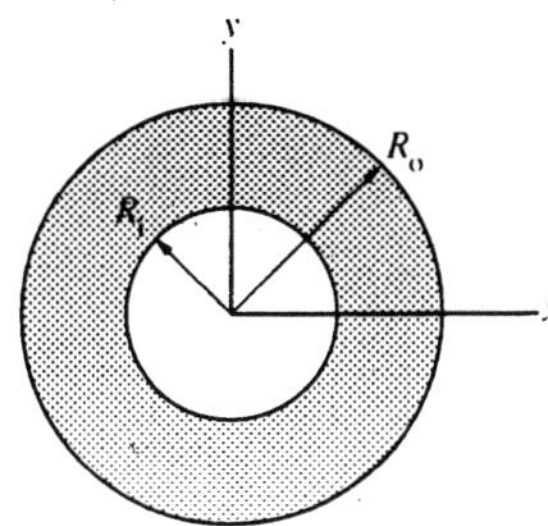

8.11ps Suppose that the thin, homogeneous 10-lb plate shown is modified by cutting a 2-in. radius circular hole centered at (7, 2.5) in. Draw the new plate to scale. Determine the weight of the modified plate, locate its center of mass, and determine its mass moments of inertia about the x, y, and z axes.

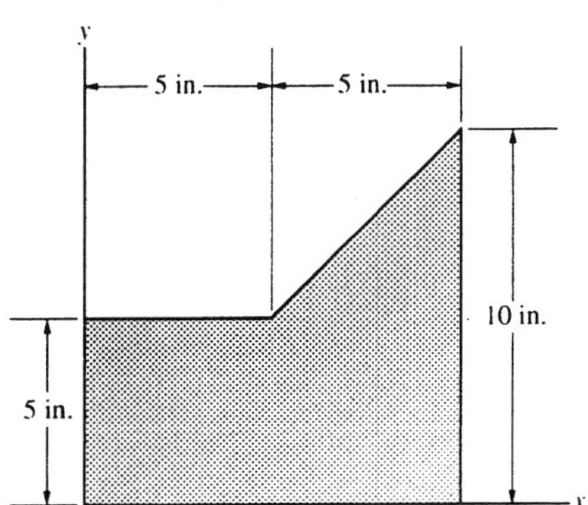

8.12ps Determine the mass moments of inertia of the modified plate described in Problem 8.11ps about x', y', and z' axes that are parallel to the axes shown but with their origin at the center of mass of the modified plate.

8.13ps Use integration to determine the mass moment of inertia of a homogeneous sphere of mass m and radius R about an axis through its center.

8.14ps A process is proposed to manufacture ball bearings in space. It results in a perfectly spherical bearing of radius R_o with a perfectly spherical void of radius R_i at its center. Let the mass moment of inertia of a solid bearing of radius R_o be I_o and let its mass be m_o. Let the mass moment of inertia and mass of a hollow bearing be I and m, respectively. Assume that all bearings have the same mass density. Draw graphs of m/m_o and I/I_o as functions of R_i/R_o for $0 \leq R_i/R_o \leq 1$.

8.15ps Locate the center of mass of the thin, homogeneous 2-kg plate and determine the mass moment of inertia about the axis perpendicular to the plate through its center of mass.

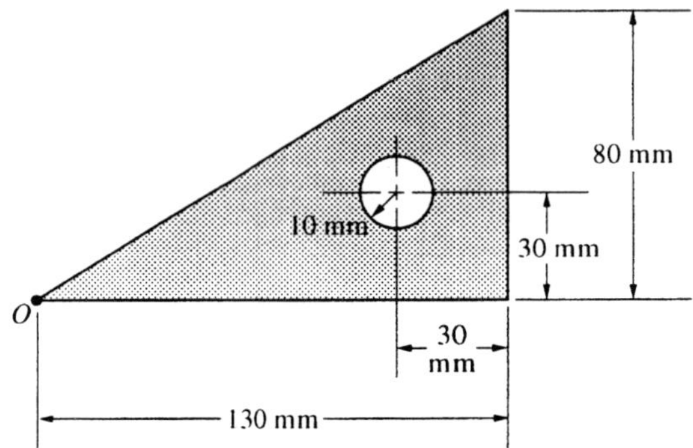

Chapter 9: Distributed Forces

9.1ps If you represent the triangular distributed load by an equivalent force, what is the magnitude of the force and where does its line of action intersect the x axis?

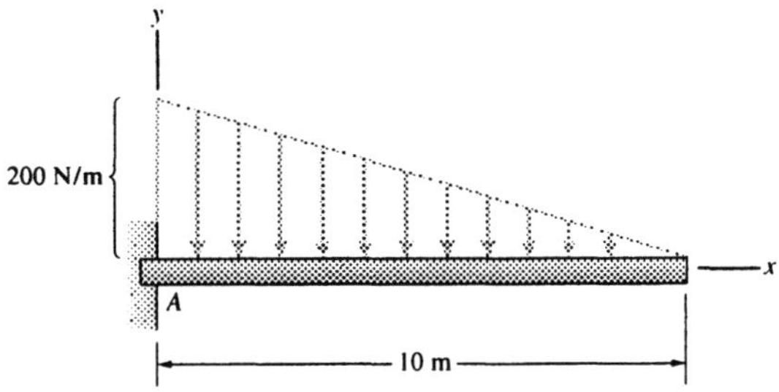

9.2ps Determine the magnitude of a single force, and the distance from A to its line of action, that results in the same reactions at A and B as the distributed load $w = 10x^2$.

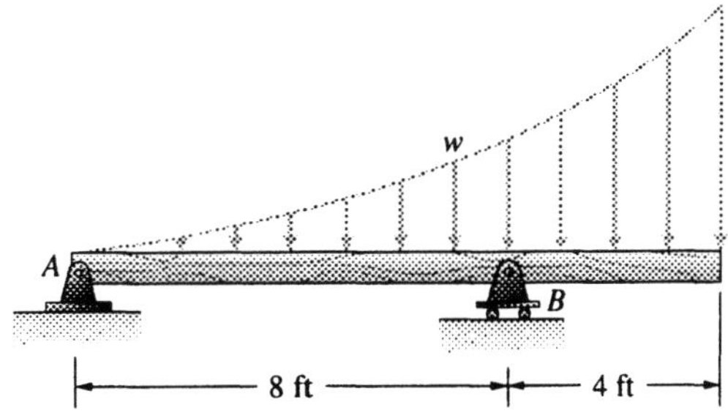

9.3ps Suppose that in addition to the 2 ft-kip couple, a 2-kip downward force is applied 0.5 ft from the left end of the beam. Determine the internal forces and moment at A, B, and C.

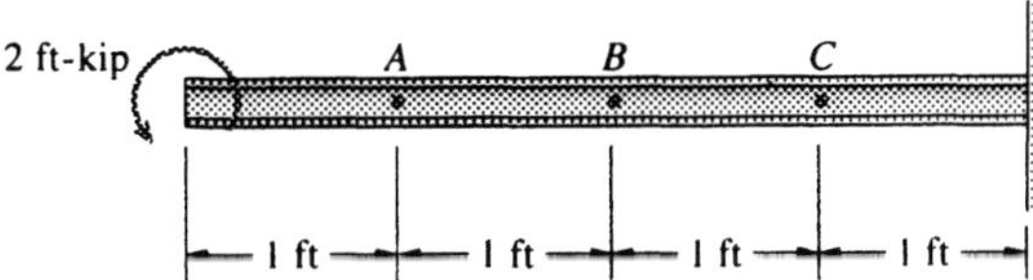

9.4ps Suppose that you are analyzing the beam shown, and your goal is to reduce the maximum positive bending moment on the beam to less than 800 N-m by adding a single point force to the beam. Find a force (and its point of application) that will meet this requirement. Draw the shear force and bending moment diagrams for your new loading.

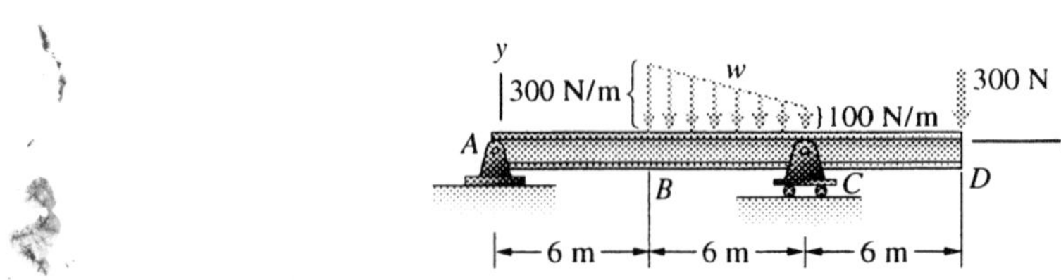

9.5ps Find the smallest possible point force and its point of application that will meet the requirement described in Problem 9.4ps.
Strategy: Set the problem up analytically and use a computer to evaluate the various possibilities.

9.6ps For the beam shown, find the magnitude and point of application of a single point force that reduces the value of the shear force at *A* to zero. Draw the free-body diagram of the beam with the new load included and draw the shear force and bending moment diagrams.

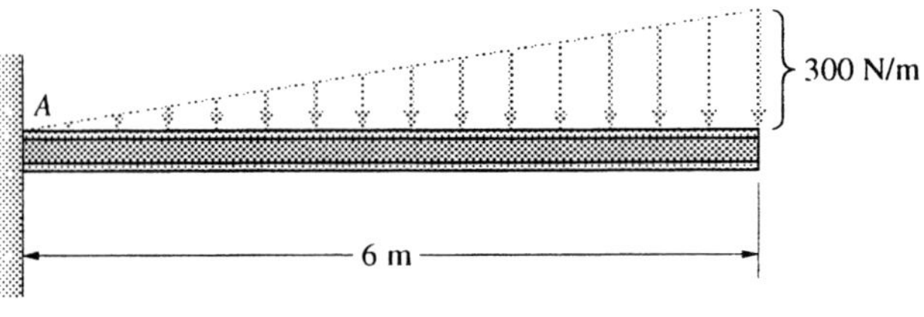

Chapter 10: Friction

10.1ps Suppose that you build an apparatus like that shown to measure coefficients of static friction by measuring the angle α at which the block A slips. Draw a graph of the angle α at which slip occurs as a function of μ_s for $0 \leq \mu_s \leq 1$.

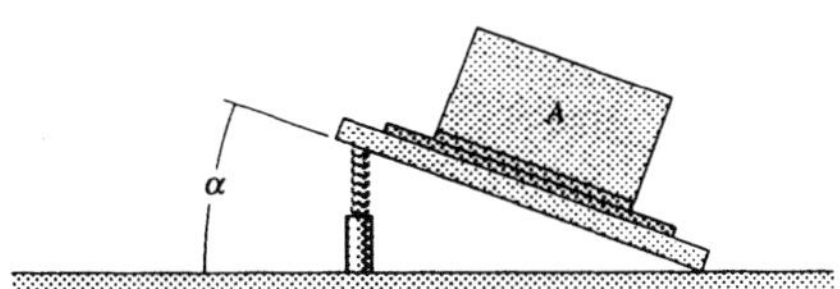

10.2ps Using the apparatus described in Problem 10.1ps as a starting point, suggest a device that could be used to measure the coefficient of kinetic friction μ_k. Describe clearly how the device would work.

10.3ps Suppose that the 20° angle is increased to 60°. The mass of the box is 5 kg. The force F is horizontal. The coefficient of static friction between the box and the surface is $\mu_s = 0.3$. Determine the range of values of F for which the box will remain stationary.

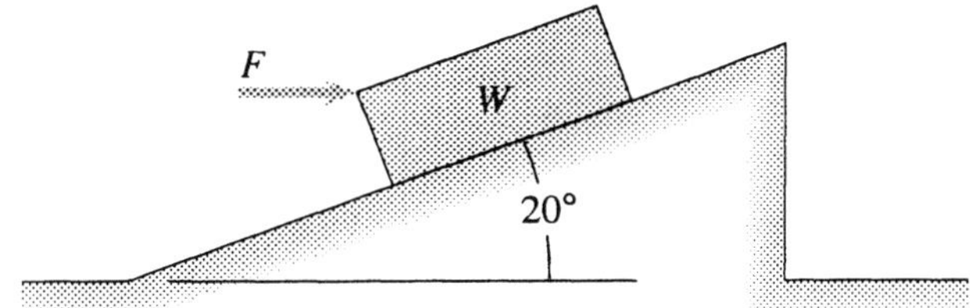

10.4ps The 20-lb homogeneous bar is 8 ft long. Neglect friction between the bar and the wall. Draw a graph of the maximum angle α for which the bar will not slip as a function of the coefficient of static friction μ_s between the bar and the floor for $0 \leq \mu_s \leq 1$.

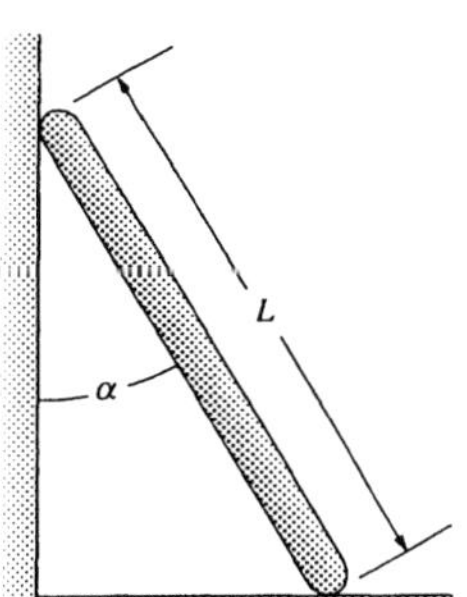

10.5ps The disk weighs 50 lb. Neglect the bar's weight. The coefficients of friction between the disk and the floor are $\mu_s = 0.6$ and $\mu_k = 0.4$. Determine the largest clockwise and counterclockwise couples that can be applied to the stationary disk without causing it to begin rotating. If the values are different, explain why.

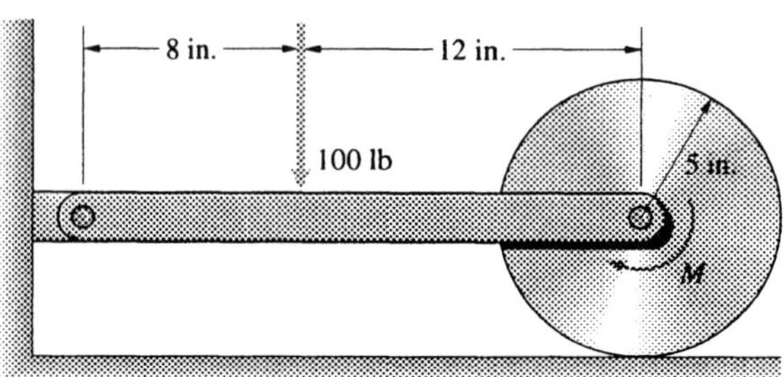

10.6ps Explain how the woman can push the stationary refrigerator in such a way that it slips and does not tip over. Also explain why the height h at which she pushes, the coefficient of friction μ_s, and the width of the refrigerator's base $2b$ are important, but the height of the refrigerator's center of mass above the floor does not matter.
Strategy: Tip is impending when the reaction at A is zero. Assume that tip and slip are impending simultaneously.

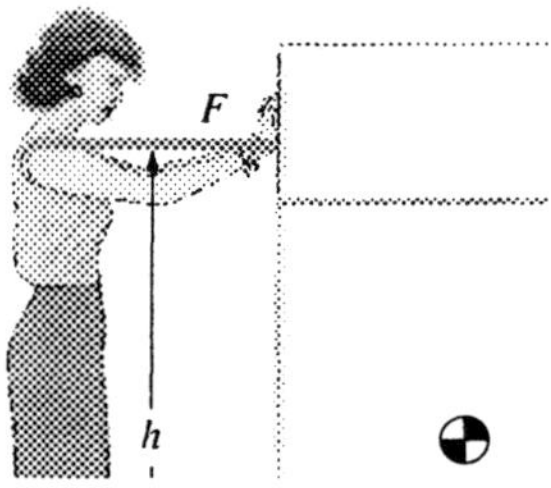

10.7ps The coefficients of friction between the wedge and log are $\mu_s = 0.3$ and $\mu_k = 0.2$. Find the range of values of α within the range $5° \leq \alpha \leq 25°$ for which the wedge will remain in place in the log when the force F is removed. Neglect the weight of the wedge.

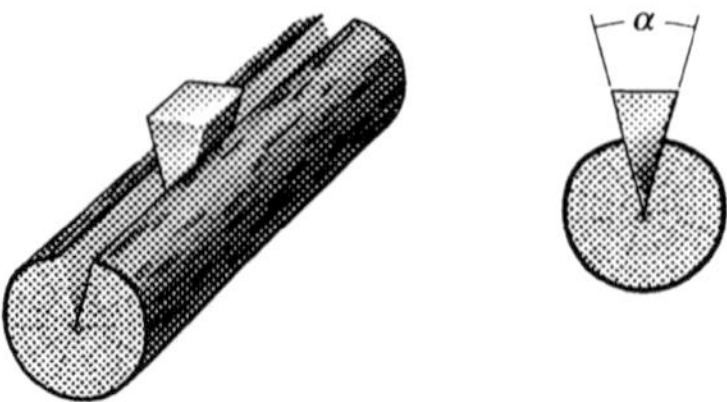

10.8ps Block *A* weighs 1,000 lb and $\mu_k = 0.2$ between all surfaces. Neglect the weight of the wedge. Determine the horizontal force *F* necessary to raise *A* at a constant rate.

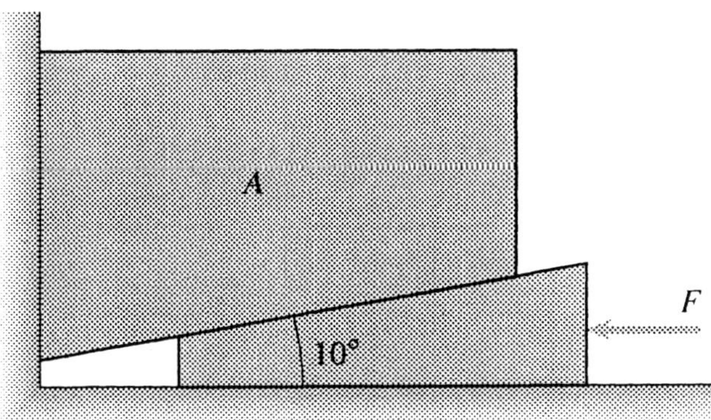

10.9ps For the system described in Problem 10.8ps, assume that the wedge angle α, which is 10° in the figure, can be varied. Draw a graph of the horizontal force *F* necessary to raise *A* at a constant rate as a function of α for $5° \le \alpha \le 30°$.

10.10ps Box *A* weighs 1,000 lb and the wedge *B* weighs 500 lb. Assume that the wedge half-angle α, which is 10° in the figure, can be varied. The coefficients of friction between all surfaces are $\mu_s = 0.3$ and $\mu_k = 0.2$. Determine the largest value of α for which the wedge *B* will remain in place with the force *F* removed.

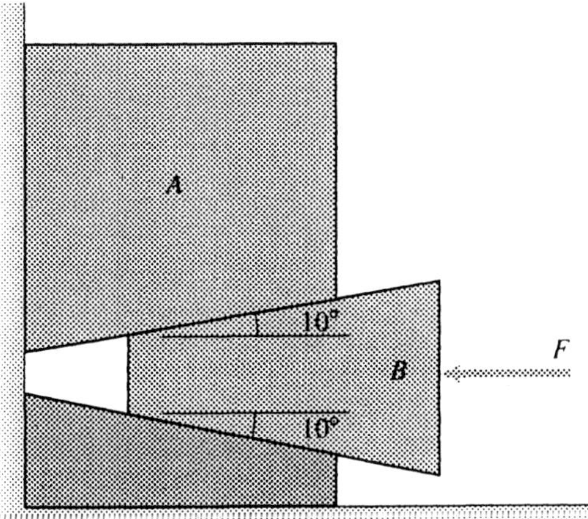

10.11ps The position of the horizontal beam can be adjusted by turning the machine screw *A*. Neglect the weight of the beam. Various screws are proposed for use in this mechanism with pitches *p* in the range $0.5 \le p \le 5$ mm. Draw a graph of the couple *C* that must be exerted on the stationary screw to rotate it and start the beam moving upward as a function of the pitch p for the stated range of values.

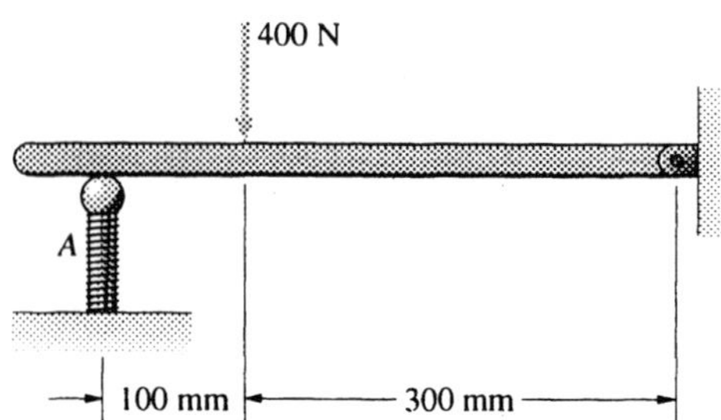

10.12ps In Problem 10.11ps, draw a graph of the couple C that must be exerted on the stationary screw to rotate it and start the beam moving downward as a function of the pitch p for the stated range of values.

10.13ps The weight of the load is $W = 1400$ lb. The pulley P has a 6-in. radius and is rigidly attached to a horizontal shaft supported by journal bearings. The radius of the shaft is 3/4 in. and the coefficient of kinetic friction between the shaft and the bearings is $\mu_k = 0.2$. Neglect the weights of the pulley and shaft. What tension must the winch exert to lower the weight at a constant rate?

10.14ps The hand sander has a rotating disk D with a four-inch radius that has sandpaper bonded to it. The coefficient of kinetic friction between the disk and the surface is $\mu_k = 0.6$. Draw a graph of the torque M necessary to turn the disk at a constant rate as a function of the downward force F exerted by the operator and the sander's weight for $5 \leq F \leq 30$ lb.

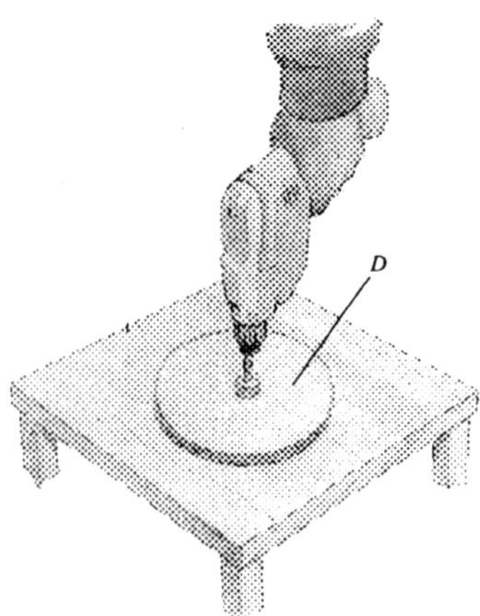

10.15ps The two pulleys are rusted and will not turn. The coefficient of static friction between the pulleys and the rope is $\mu_s = 0.6$. What is the smallest tension T necessary to hold the 180-lb man stationary?

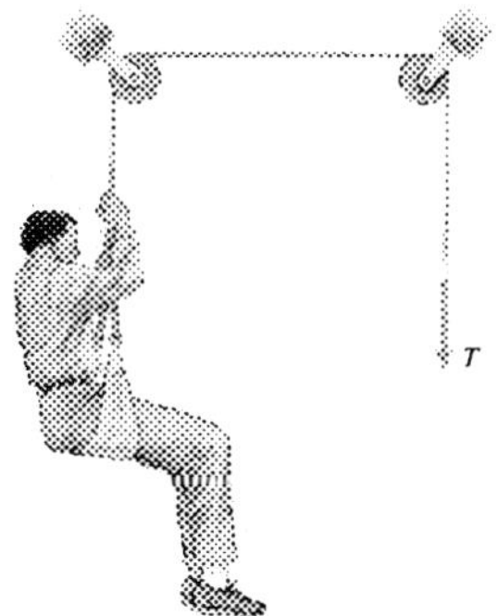

10.16ps The student wants to use the tree limb to hold a 500-lb weight suspended above the ground. He considers three methods. Simply passing the rope over the limb as shown, or giving the rope one additional wrap around the limb, or giving it two wraps. He determines that with the three methods the arcs over which the rope contacts the limb are 130°, 490°, and 850°. The coefficient of static friction between the rope and the tree limb is 0.5. Determine the minimum force he must exert to hold the weight suspended above the ground for each case.

Chapter 11: Virtual Work and Potential Energy

11.1ps Suppose that in addition to the loading shown, a 200-lb downward force acts at *B*. Use the principle of virtual work to determine the horizontal reaction at *A*.

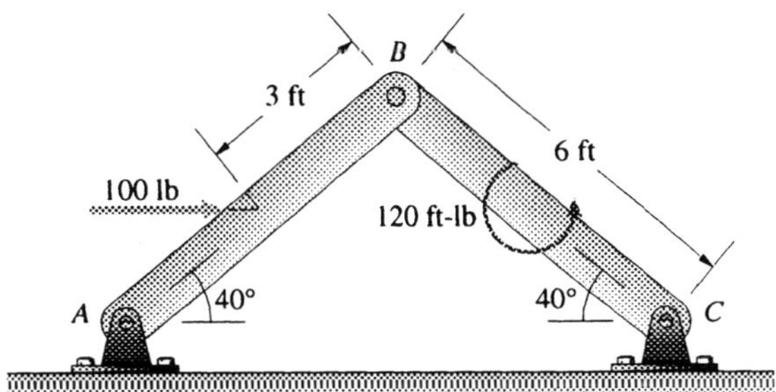

Dynamics

Chapter 1: Introduction

1.1ps The acceleration due to gravity on the surface of the Earth is 32.2 ft/s^2 and on the surface of Mars it is 13.2 ft/s^2. Suppose that a supply ship delivers a cargo to a future Mars colony that weighed 3780 lb on the surface of the Earth. What is its weight on the surface of Mars?

1.2ps The acceleration due to gravity on the surface of the Earth is 32.2 ft/s^2 and on the surface of the Moon it is 5.32 ft/s^2. Suppose that a future geologist working at the Farside Lunar Base sends a sample back to a laboratory on the Earth. The sample weighed 25.2 lb. on the Moon. Determine the weight of the sample on the surface of the Earth.

1.3ps The radius of the Earth is 6370 km and its mass is 5.97×10^{24} kg. The acceleration due to gravity at the surface of the Earth is 32.2 ft/s^2. The radius of the Moon is 1738 km and the acceleration due to gravity at the surface of the Moon is 5.32 ft/s^2. Convert these quantities to a common system of units and use them to determine the mass of the Moon in kg.

1.4ps On its way from the Earth to the Moon, a spacecraft travels at a speed of 37,951 ft/s. What is the speed of the spacecraft in m/s?

1.5ps The mass of a cubic meter of water is 1000 kg. If $g = 9.81$ m/s^2, determine the weight of one cubic meter of water in newtons and in pounds.

1.6ps Water at the surface of the Earth weighs 62.4 lb/ft^3. Determine the volume of 50,000 lb of water. If it were formed into a cube, what would be the length of each side of the cube?

1.7ps A compact car has a top speed of 93.2 miles per hour. Determine its top speed in ft/s, km/hr, and m/s.

1.8ps Race cars on the Indianapolis Speedway travel at 220 miles per hour on the straight sections of the track. Express this speed in m/s, ft/s, and km/hr.

1.9ps The average weight density of the human body is approximately the same as that of water, 62.4 lb/ft^3. Draw graphs of weight in pounds versus volume in cubic feet and mass in kilograms versus volume in cubic meters for the average human. Assuming that you are of average density, find and mark your own values on the graphs.

1.10ps The speed of light in vacuum is 299,792.5 km/s. Express this speed in miles per hour, miles per second, and km per hour.

1.11ps A particular glacier moves at a speed of 2.5 meters per year. The average length of a year, taking leap years into consideration, is 365.25 days. Express the speed of the glacier in miles per hour, inches per day, meters per second, and yards per week.

1.12ps The acceleration due to gravity at the surface of the Moon is 1.62 m/s^2 and the radius of the Moon is 1738 km. Draw a graph of the weight of a 1-kg object (the gravitational force exerted on the object by the Moon) as a function of the object's altitude above the surface of the Moon for altitudes from zero to 10,000 km.

1.13ps Calculate your age as of midnight last December 31. Ignore time zone effects. If you know the time of day that you were born, use that time. Otherwise, assume 7 AM as the time you were born. Express your age in days and fractions of a day, hours and fractions of an hour, minutes and fractions of a minute, and finally, seconds and fractions of a second. (Be sure to take leap years into account.)

Chapter 2: Motion of a Point

2.1ps The train starts from rest and has a constant acceleration of 0.5 ft/s^2 until its speed reaches 120 ft/s. After this time its speed is constant. How long from the time the train started from rest does it take to travel five miles?

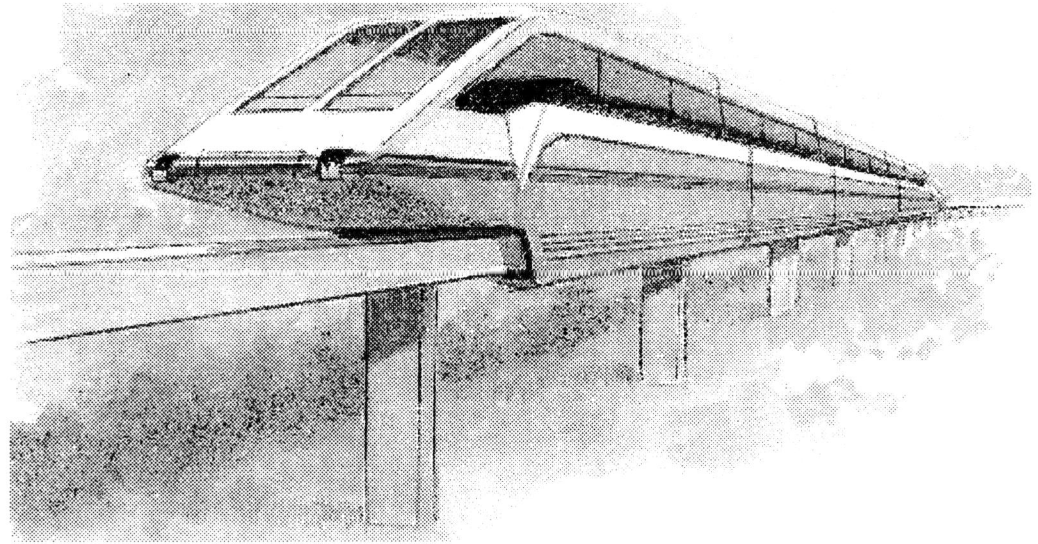

2.2ps The position of a point is given by $s = 6t^3 - 4t^2 + 2t - 1$ ft. Determine: (a) the displacement of the point during the interval from $t = 2$ seconds to $t = 7$ seconds; (b) the velocity of the point at $t = 3$ seconds; (c) the acceleration of the point at $t = 2$ seconds.

2.3ps The position of a point during the time interval $0 \leq t \leq 10$ seconds is given by $x = -3t^4 + 6t^3 - 3t^2 + 6t$ meters.
(a) Draw graphs of the point's position, velocity, and acceleration as functions of time over the specified interval.
(b) What do you notice about the graph of the position at times when the velocity is zero?
(c) What do you notice about the graph of the velocity at times when the acceleration is zero?
(d) Find the time when the absolute value of the velocity is a maximum and the time when the absolute value of the acceleration is a maximum. Mark these points on your graphs.

2.4ps A steel ball dropped into the ocean has a downward acceleration given by $a = 0.9g - cv$, where $c = 3.02$ s^{-1}, $g = 32.2$ ft/s^2, and v is the ball's velocity in ft/s. Draw graphs of the ball's velocity and the depth it has reached as functions of time for $0 \leq t \leq 70$ minutes.

2.5ps A ball is thrown from the top of a building with an initial speed of 50 ft/s.

Suppose that instead of being thrown horizontally as shown, it is thrown at a 40° angle above the horizontal.
(a) What maximum height above the ground does the ball reach?
(b) At what horizontal distance from the base of the building does the ball strike the ground?
(c) How long is the ball in the air?

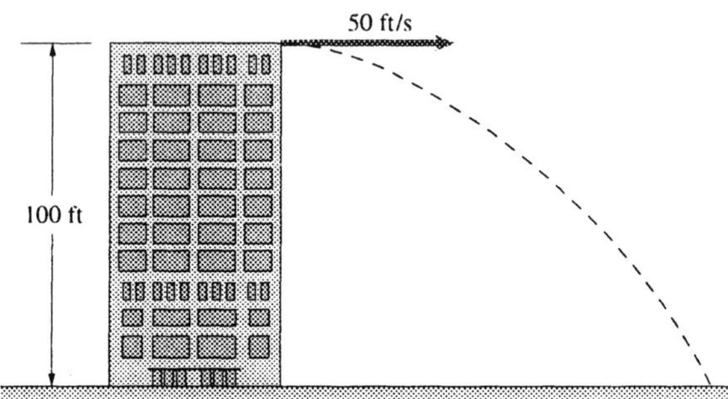

2.6ps A pilot wants to drop supplies to a remote location. He plans to release the package while in level flight at a height $h = 200$ meters and a speed $v_0 = 70$ m/s. Determine the horizontal distance d from the release point to the target.

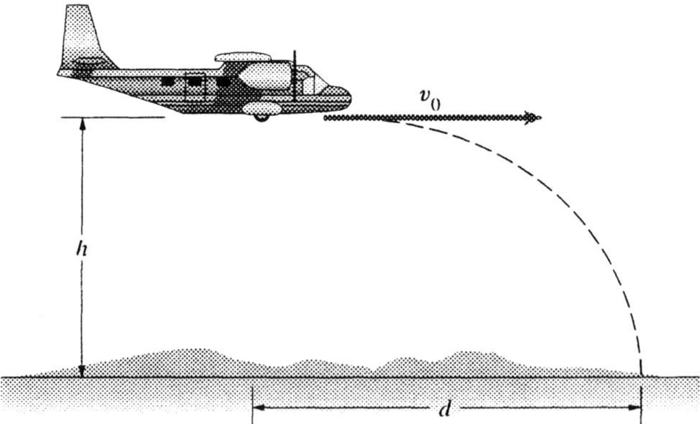

2.7ps Suppose that the pilot in Problem 2.6ps releases the package at the correct point with the correct altitude of 300 ft and correct speed of 200 ft/s, but is inadvertently climbing at an angle of 3° relative to the horizontal instead of flying level. How far does he miss his target?

2.8ps Suppose that the pilot in Problem 2.6ps releases the package while in level flight at the correct altitude and speed, but releases the package .2 seconds late. How far does he miss his target?

2.9ps The batter hits the baseball when it is 2.5 ft directly above the center of home plate. The ball leaves the bat at an angle of 35° above the horizontal at

120 ft/s. How far from home plate does the ball strike the ground? Neglect drag.

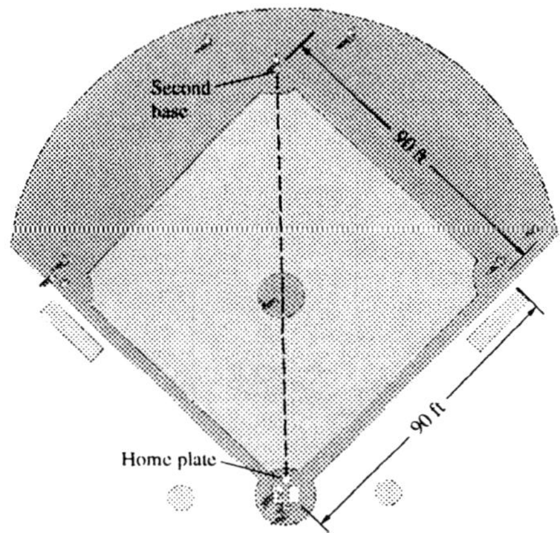

2.10ps Assume that the batter in Problem 2.9ps hits the ball at an angle of 40° above the horizontal at a speed v. Draw a graph of the distance from home plate that the ball strikes the ground for $20 \le v \le 120$ ft/s.

2.11ps Instead of attempting to enter the water 6 ft out from the base of the cliff as shown in the figure, the diver runs to the takeoff point and dives out horizontally at 25 ft/s. He strikes the crest of a wave 85.5 ft below his takeoff point. How far out from the cliff base does he strike the water?

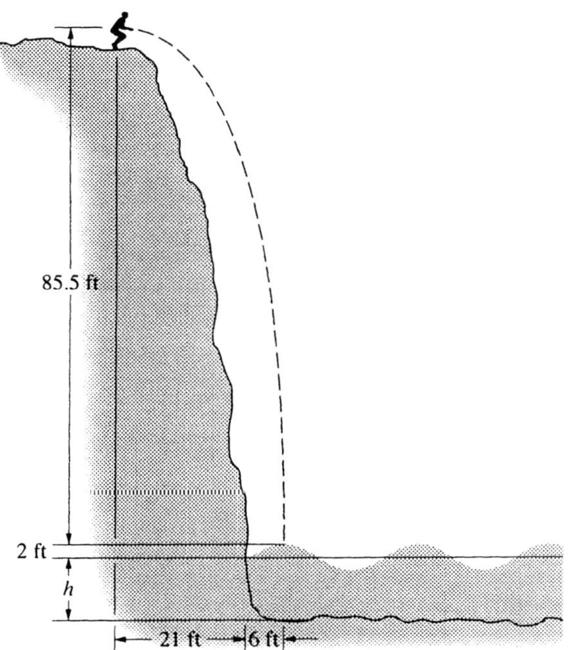

2.12ps A jet engine is rotating at 5,000 rpm when the throttle is advanced. The angular acceleration of the engine is given by $\alpha = C \sin(\pi t/10)$ rad/s^2. The engine reaches 10,000 rpm 5 seconds after the throttle is advanced. What is the constant C?

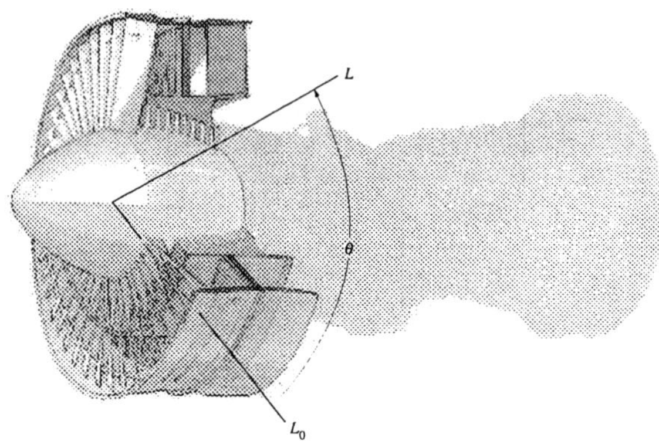

2.13ps Suppose that you are assigned to recalibrate the clock to work on a Mars colony. You want to have 24' "Mars hours" in a "Mars day" and 60 "Mars minutes" in a "Mars hour". The duration of a Mars day in Earth time is 24 hours and 36 minutes. Determine the necessary angular velocities of the hour hand and minute hand for the "Mars clock" in degrees/Earth second.

2.14ps Using the data given in Problem 2.13ps, determine the angular velocity of the planet Mars in rad/s and degree/s.

2.15ps The angular velocity of the Earth is slowly decreasing due to tidal interaction with the Moon. Records indicate that at one time in the past there were 430 days in a year. Assume that the year (the time for the Earth to make one orbit around the Sun) has not changed. At present there are 365.25 days in a year. Determine the angular velocity of the Earth at present and also at the time that there were 430 days in a year. Assume, for this problem, that the Earth turns 360 degrees in 24 hours.

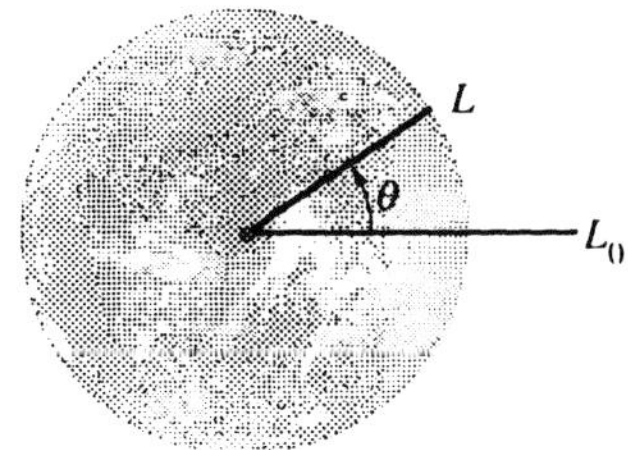

2.16ps A rotor, turning at 100 rpm at $t = 0$, is subjected to the angular acceleration $\alpha = 5\sin(\pi t/10)$ rad/s^2 for 25 seconds. What is its angular velocity at $t = 25$s?

2.17ps The motorcycle rider starts from rest at $t = 0$ on the circular track of 400-m radius. Her tangential acceleration is given by $a_t = 2 + 0.1t$ m/s^2. If the maximum total acceleration that the motorcycle can experience without losing traction and "spinning out" is 5.90 m/s^2, determine the time at which the rider must stop increasing her speed. What is the maximum speed that the motorcycle can reach on this track without losing traction?

2.18ps A satellite is in a circular orbit of radius R around the Earth. Use $g = 9.81$ m/s^2 and $R_E = 6378.14$ km. Draw a graph of the orbital velocity as a function of R for $6{,}550 \le R \le 400{,}000$ km. The Moon's orbit around the Earth is circular with a radius of 384,000 km. Use your graph to determine the Moon's orbital velocity.

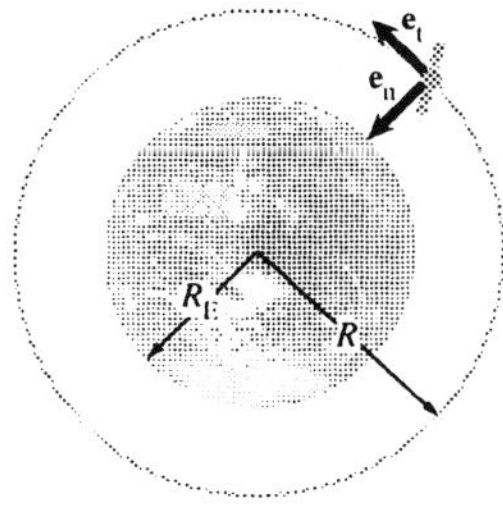

2.19ps Problem 2.18ps involved drawing a graph of the velocity of a satellite in a circular orbit around the Earth as a function of the radius of the orbit. Draw a graph of the time required for one orbit (the period) as a function of R for $6{,}550 \le R \le 400{,}000$ km. Use your graph to determine the Moon's period in days.

2.20ps Suppose that a car passes point B moving toward the left at 40 mph. It has constant acceleration from B to A, reaching 60 mph at point A. Determine the magnitude of the car's total acceleration: (a) at a point slightly more than 80 ft from B; (b) just before it reaches the "midpoint" of the hill (where the two 30° arcs meet); (c) when it is slightly more than 80 ft from A; (d) when it is slightly less than 80 ft from A.

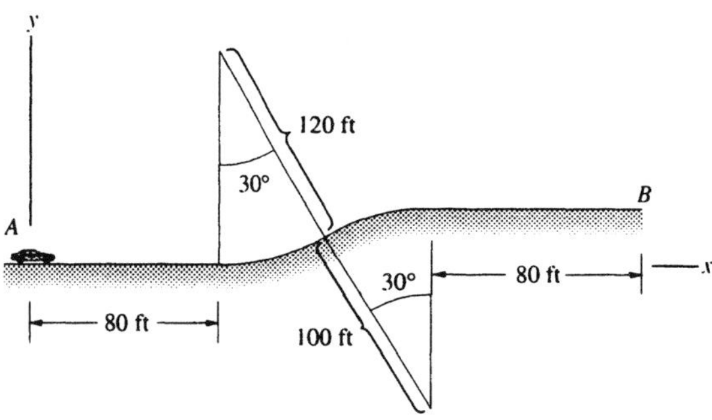

2.21ps The centrifuge is rotating at a constant angular velocity so that the magnitude of the total acceleration the pilot trainee inside is subjected to (including the acceleration of gravity) is ten g's. Determine the angular velocity of the centrifuge.

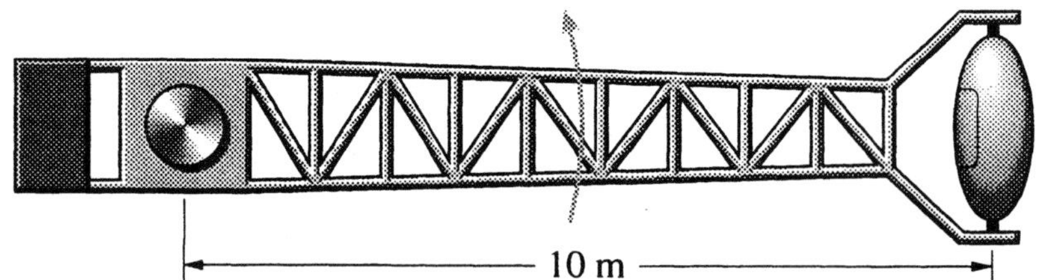

2.22ps The centrifuge shown in Problem 2.21ps is moving at a constant angular velocity, subjecting the pilot trainee to a total acceleration of ten-g magnitude, when an emergency occurs requiring that the centrifuge be stopped quickly. Draw a graph of the maximum allowable angular deceleration as a function of the angular velocity as the centrifuge slows down if the trainee can withstand a total acceleration magnitude of fourteen g's.

2.23ps The astronaut training aircraft shown actually enters the "weightless" segment of its trajectory while climbing at an angle 60° above the horizontal. The aircraft follows the weightless path until it is descending at 60° below the horizontal. If the weightless segment of the flight is 25 seconds long, what is the airplane's velocity at the beginning of the weightless segment?

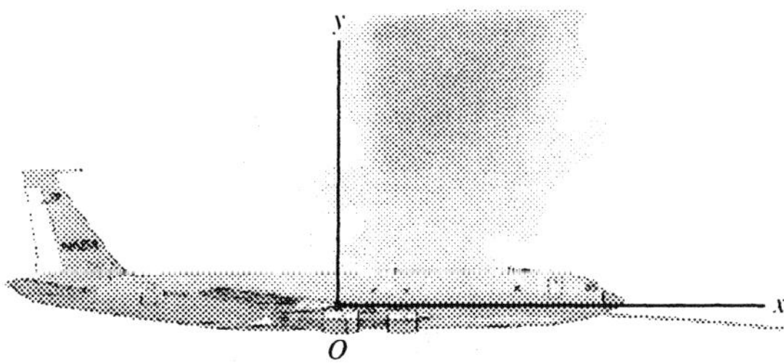

2.24ps The astronaut training aircraft described in Problem 2.22ps enters the "weightless" segment of its trajectory at an altitude of 25,500 ft. Determine the maximum altitude reached by the aircraft and its speed at the maximum altitude. Draw the free-body diagram of the aircraft when it is at its maximum altitude.

2.25ps The disk's angular position is given by $\theta = 0.1t^2$ rad. Suppose that at $t = 0$, you begin walking along a radial line painted on the disk in such a way that your radial position is $r = 0.5t^2 + t$ (ft). Determine your position, velocity, and acceleration in terms of polar coordinates at $t = 3$ seconds.

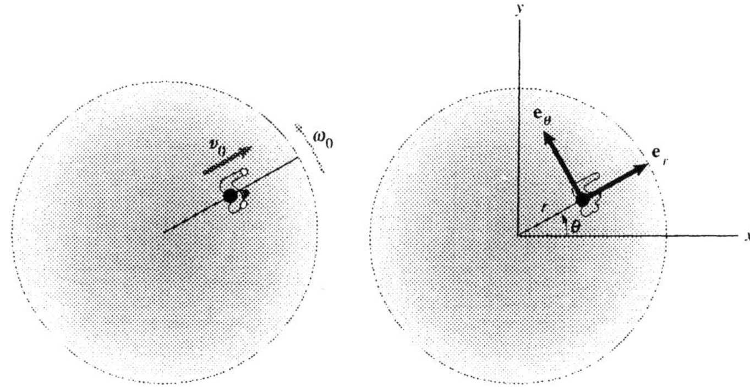

2.26ps The polar coordinates of a point are

$$r = 9t^4 - 6t \text{ meters},$$
$$\theta = 2t^3 \quad 7t^2 \text{ radians}$$

Determine the velocity and acceleration of the point in terms of polar coordinates at $t = 2$ s.

2.27ps For the point whose motion is described in Problem 2.26ps, determine the velocity and acceleration in terms of cartesian coordinates at $t = 2$ s.

2.28ps For the point whose motion is described in Problem 2.26ps, determine the velocity and acceleration in terms of normal and tangential components at $t = 2$s.

2.29ps Consider the point whose motion is described in Problem 2.26ps. Draw graphs of the x and y coordinates of the point $0 \le t \le 5$ seconds. Draw graphs of the x and y components of the velocity for the same time interval.

2.30ps The P-51 Mustang is flying at 400 miles per hour at the bottom of a loop of 2,000-ft radius and is accelerating along its path at 5 ft/s^2. Determine the magnitude of the total acceleration of the aircraft when it is at the bottom of the loop.

2.31ps A transfer orbit used to take communications satellites from low Earth orbit to geosynchronous orbit has a perigee radius of 6,673 km and an apogee radius of 42,164.2 km. What is the eccentricity of the orbit?

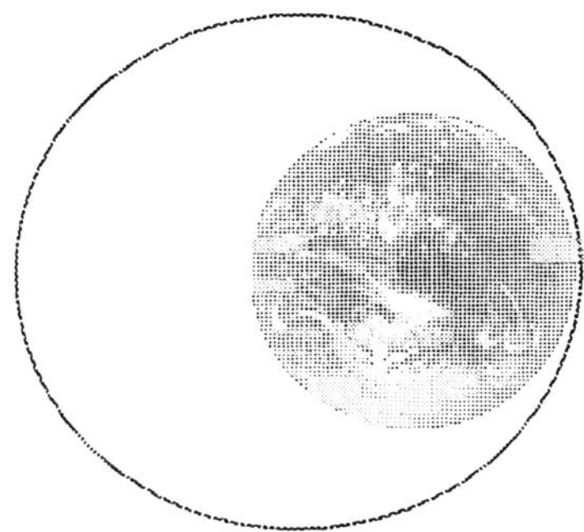

2.32ps For the transfer orbit described in Problem 2.31ps, determine the velocities at perigee and apogee.

2.33ps The initial part of the actual trajectory used to send spacecraft from the Earth to the Moon can be approximated by neglecting the effect of the Moon's gravity and assuming an elliptic orbit with perigee near the Earth (assume a perigee radius of 6,670 km) and an apogee equal to the radius of the Moon's orbit around the Earth, approximately 384,000 km. Determine the eccentricity of the transfer orbit and the velocity at perigee. Refer to above figure from problem 2.31ps.

2.34ps The acceleration due to gravity at the surface of the Moon is 1.62 m/s^2 and the Moon's radius is 1,738 km. Determine the velocity of a spacecraft in a circular orbit around the Moon with a radius of 1,838 km. What is the period of the orbit?

2.35ps Example 2.14 on pages 79–81 of the text contains a simplifying approximation. The period of a geosynchronous orbit (the time required for the Earth to undergo one rotation, or 360°) is assumed to be exactly 24 hours. In reality the Earth rotates more than 360° in 24 hours. This is due to the Earth's motion in its orbit around the Sun. The Earth rotates about its axis in the same direction it rotates about the Sun. As a result, to get from noon to noon (24 hours) at a fixed location the Earth must rotate, not 360°, but almost 361°. The actual time required for the Earth to rotate 360° is 23 hours, 56 minutes, 4.09 seconds. Using this information, determine the true radius of a geosynchronous orbit and the true velocity of a satellite in geosynchronous orbit.

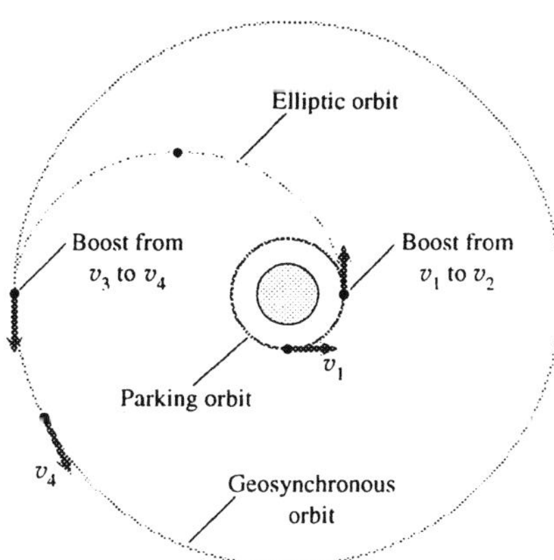

2.36ps Airline pilots try to choose altitudes to take advantage of favorable winds and minimize the fuel consumption and time penalty caused by unfavorable winds. Assume that a jet stream wind is flowing toward the east at 100 mi/hr and an airliner is flying at an airspeed (the magnitude of its velocity relative to the air) of 525 mi/hr. What direction should the pilot fly relative to the air if he wants his path relative to the ground to point 30° east of north? What will be the aircraft's ground speed (the magnitude of its velocity relative to the ground)?

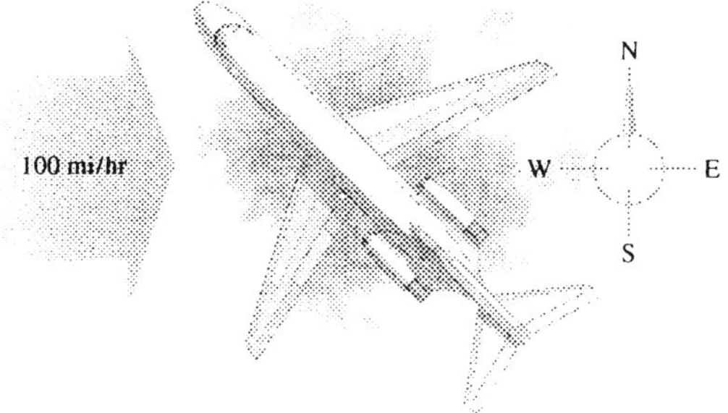

2.37ps Assume that the jet stream wind shown in Problem 2.36ps is flowing at 100 mi/hr toward the east and an airliner is flying at an airspeed of 525 mi/hr. The direction the aircraft points is called its heading. The direction of its velocity vector relative to the ground is its course. The direction and course are measured in degrees clockwise from north. For example, if the aircraft points east, its heading is 90°. Draw graphs of the heading and ground speed as functions of the course for course directions from 0° to 360°.

2.38ps The point P moves along a spiral path defined by

$$r = 0.5\theta^2 \text{ meters,}$$
$$\theta = 2t + \sin(t) \text{ radians.}$$

Determine the position, velocity, and acceleration of the point in terms of polar coordinates at $t = 5$ seconds.

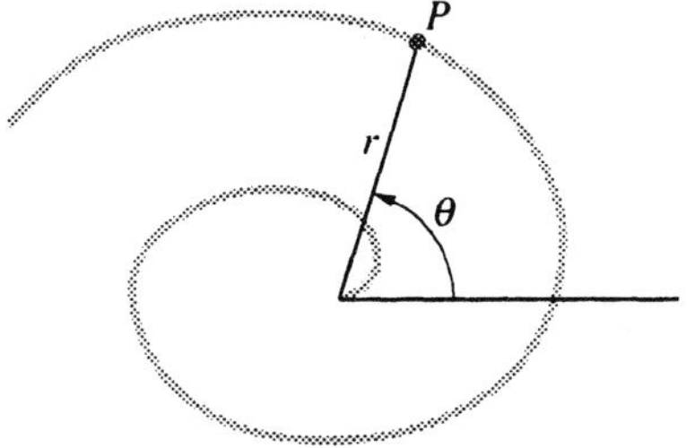

Chapter 3: Force, Mass, and Acceleration

3.1ps The total external force acting on a 100-kg object is $F = 100\mathbf{i} - 73\mathbf{j} + 60\mathbf{k}$ (N). What is the acceleration of the object's center of mass? What is the magnitude of its acceleration?

3.2ps The position of the center of mass of a 150-lb object is given by $r = t^2/2\mathbf{i} - 6t\mathbf{j} + \sin(\pi t)\,\mathbf{k}$ (ft). Determine the total external force acting on the object at $t = 0.5$s and at $t = 2$s.

3.3ps The 2-lb collar A is initially at rest on the smooth horizontal bar. At $t = 0$, the force $\mathbf{F} = Ct^2\mathbf{i} + 6t\mathbf{j} - 3\mathbf{k}$ (lb) is applied to the collar. Determine the constant C if the collar is to reach the right end of the bar with a speed of 10 ft/s.

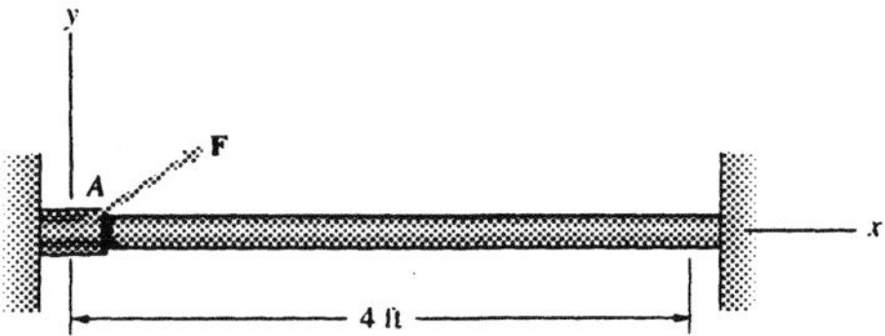

3.4ps Suppose a 170-lb person is standing on scales that measure his weight in an elevator that is moving upward at a constant speed of 20 ft/s. If a malfunction causes the elevator's emergency brakes to bring the elevator to a stop with a constant deceleration over a distance of 3 ft, what do the scales read during the deceleration?

3.5ps In Problem 3.4ps, what do the scales read during the deceleration if the elevator is moving downward at 20 ft/s?

3.6ps The airplane weighs 16,000 lb. At the instant shown its thrust is $T = 12{,}000$ lb, the aerodynamic lift force perpendicular to the runway is 8,000 lb, and the aerodynamic drag force parallel to the runway is 4,000 lb. What is the airplane's acceleration?

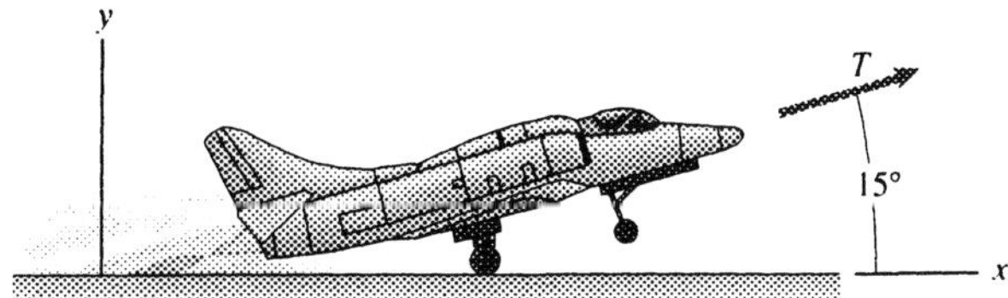

3.7ps A perfectly released 44-lb curling stone slides 31 yards from the point of release, stopping at the center of the target area. If $\mu_k = 0.01$, and if the stone is released on-line at a speed which is 0.01 ft/s too fast, how far past the center of the target area does the stone slide?

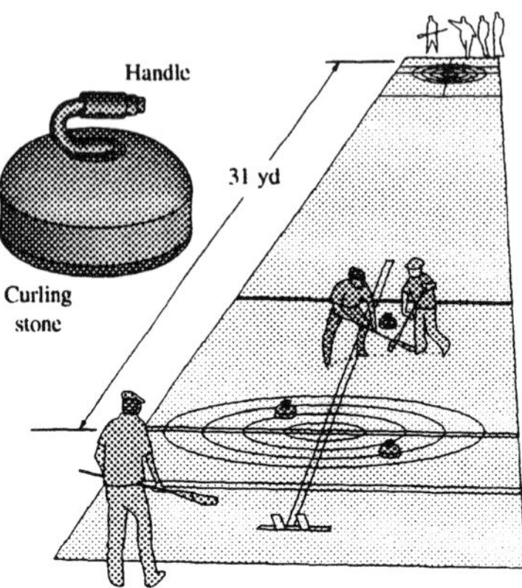

3.8ps A perfectly released 44-lb curling stone slides 31 yards from the point of release, stopping at the center of the target area. (See Problem 3.7ps.) The coefficient of kinetic friction $\mu_k = 0.01$. Find the miss distance for an on-line release that has a speed 0.99 times the velocity required for a perfect release.

3.9ps Suppose that the curling stone described in Problem 3.7ps is released with perfect speed but the angle is 0.1° to the left of the correct direction. Determine the miss distance.

3.10ps The weights weigh 10 lb and 50 lb at sea level on Earth. Suppose that the apparatus is moved to the surface of the Moon where the acceleration due to gravity is 5.32 ft/s^2. If the weights are released from rest, how far does the right weight fall in the first 0.5 seconds?

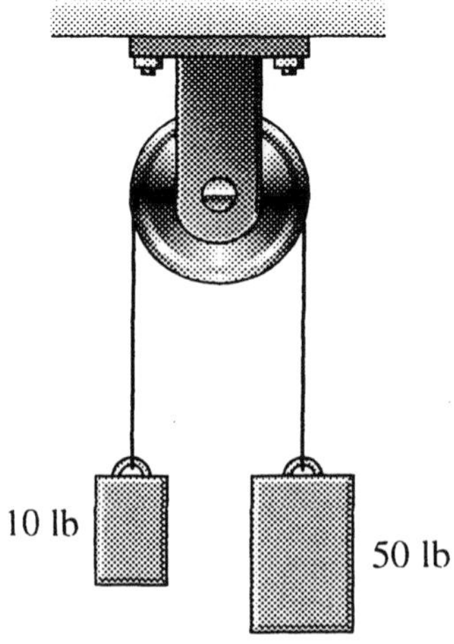

3.11ps In Problem 3.10ps, compare the ratio of the distances traveled in the first 0.5 seconds by the right weight on the Earth and on the Moon to the ratio of the acceleration due to gravity on the Earth and the Moon.

3.12ps The system shown in Problem 3.10ps is set up on the Earth (the Earth weights are shown in the figure) and also on the Moon. Find the times required for the right weight to fall a distance of one foot starting from rest in each place. How is the ratio of these times related to the ratio of the gravitational accelerations?

3.13ps The crate weighs 100 lb and the kinetic coefficient of friction between the crate and the surface is $\mu_k = 0.2$. Suppose that instead of being horizontal as shown, the force F points 10° above the horizontal. Find the value of F that will cause the crate to accelerate up the plane at 5 m/s^2.

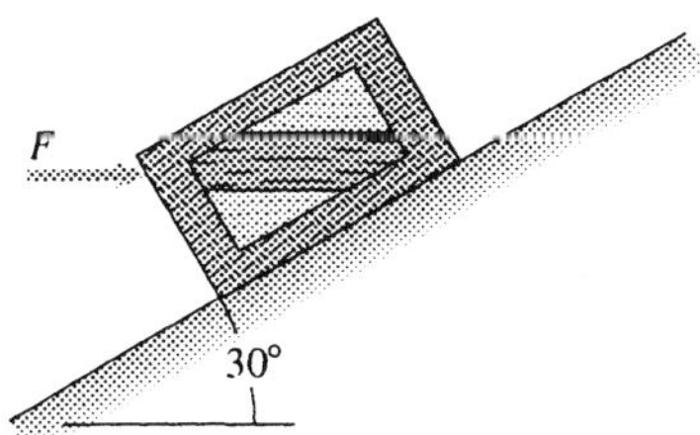

3.14ps The sky diver and parachute weigh 180 lb and are falling straight downward at 170 ft/s when the parachute opens. The magnitude of the drag force with the parachute open is $0.07v^2$.
(a) Determine the maximum deceleration experienced by the sky diver.
(b) Draw a graph of the speed of the sky diver as a function of distance d measured from the point where the parachute opens for $0 \leq d \leq 100$ ft.

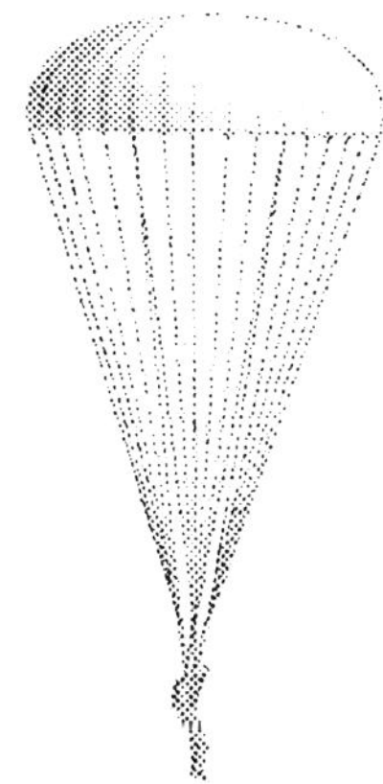

3.15ps Suppose that you are a design engineer working on a preliminary design for a rotating space station. Let R be the radial distance from the center of the rotating station to the occupied outer rim. If you want the occupants to be subjected to the equivalent of 40% of their weight on Earth at sea level, draw a graph of the angular velocity the station must have as a function of R for $60 \leq R \leq 200$ m.

3.16ps As a design engineer for the magnetically levitated railway shown, your task is to determine the bank angle of the track so that the side force S exerted on the train is zero at the nominal train speed. (The bank angle shown is 40°.) For a track segment with a turn radius of 200 m, draw a graph of the track bank angle as a function of the train speed v for $10 \leq v \leq 50$ m/s.

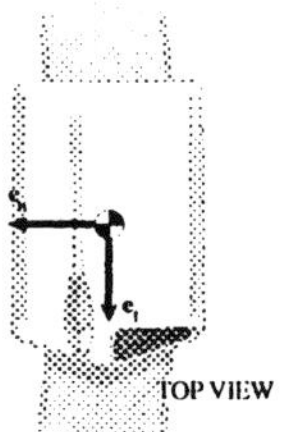

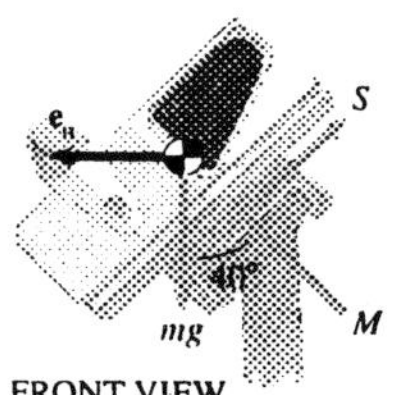

3.17ps The acceleration due to gravity at a distance R from the center of the Earth is given by gR^2_E/R^2 where $g = 9.81$ m/s^2 and the radius of the Earth is $R_E = 6.376 \times 10^6$ m. Find the velocity of an Earth satellite that is in a circular orbit of 12,000-km radius. If the gravitational force on the satellite is 1,280 N, what is its mass?

3.18ps Suppose that you work for a highway design firm and are developing graphs to show the relationship between the radius R of a circular segment of highway, the coefficient of static friction μ_s, and the maximum speed v at which it can be negotiated by a car without its tires skidding. The roadway turn is not banked. Draw graphs of the maximum speed as a function of the curve radius over the range $30 \leq R \leq 1000$ m for values of the friction coefficient μ_s=0.1, 0.3, 0.5, and 0.7.

3.19ps The load factor of an aircraft is defined to be the ratio of the lift force L to the weight W of the aircraft. The aircraft shown weighs 180,000 lb and is making a turn at constant altitude with a bank angle of 15°.
(a) Determine the load factor and the radius of curvature of the airplane's turn in the 15° banked turn shown if the airplane is flying at a constant speed of 400 ft/s.

(b) If the airplane's speed in the 15° banked turn is 800 ft/s, are the load factor and radius of curvature different from the values you obtained in part (a)? If so, what are they?

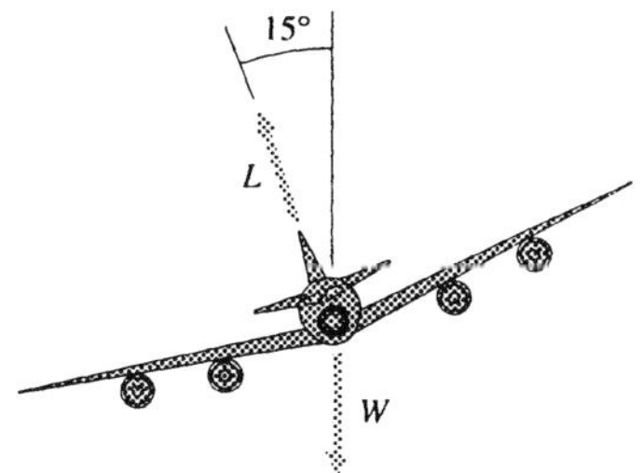

3.20ps Problem 3.19ps demonstrated that the load factor L/W of the turning aircraft is independent of the airplane's speed. Draw a graph of the load factor as a function of the bank angle β for $0 \le \beta \le 85°$.

3.21ps A car is traveling at 65 mi/hr on a straight level road when the driver sees a hazard, applies the brakes, and brings the car to a stop. The coefficients of friction between the wheels and road are $\mu_s = 0.6$ and $\mu_k = 0.4$. Assuming that he has a reaction time of 0.4 second (to start braking after the hazard is recognized), how far does the car travel from the time he sees the hazard if: (a) the driver locks all four wheels, causing the car to slide to a stop; (b) the car has anti-lock brakes which apply the maximum possible braking force without allowing the wheels to slide.

3.22ps For cases (a) and (b) in Problem 3.21ps, draw graphs of the distance the car travels (in feet) as a function of the car's initial velocity v for $55 \le v \le$ 85 mi/hr.

3.23ps If the hill is circular with radius $R = 200$ m, how fast must the car be traveling to lose contact with the road at the top of the hill?
Strategy: Draw a free-body diagram of the car and express Newton's second law in terms of normal and tangential components.

3.24ps Consider the car and hill shown in Problem 3.23ps. A stunt driver asks you to produce a graph of the maximum speed (in mi/hr) at which he can drive over the hill without losing contact with the road as a function of the radius of curvature R for $100 \le R \le 2{,}000$ ft.

3.25ps A particular car bottoms out (parts of its undercarriage come into contact with the road) when it is subjected to a load factor (the ratio of the normal force exerted on the car by the road to the weight of the car) of 2.5. How fast can the car go through a circular depression with a 500-ft radius without bottoming out?

3.26ps The robot is programmed so that the 0.4-kg part A describes the path

$$r = 1 - 0.5\cos(2\pi t) \text{ meters,}$$
$$\theta = 0.5 - 0.2\cos(2\pi t) \text{ radians.}$$

At $t = 1$ second, determine the magnitude of the force exerted on part A by the robot's jaws.

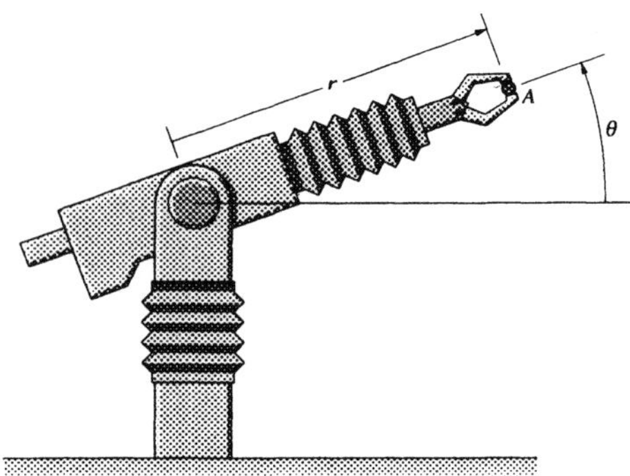

3.27ps In Problem 3.26ps, draw a graph of the magnitude of the force exerted on part A by the robot's jaws for $0 \leq t \leq 1$ seconds.

3.28ps The ski jumper passes point A moving at 17 m/s. The radius of curvature of the path from A to B is 6 m. Determine the maximum load factor (load factor = normal force/weight) he experiences on the curved part of the ramp. Neglect friction and aerodynamic drag.

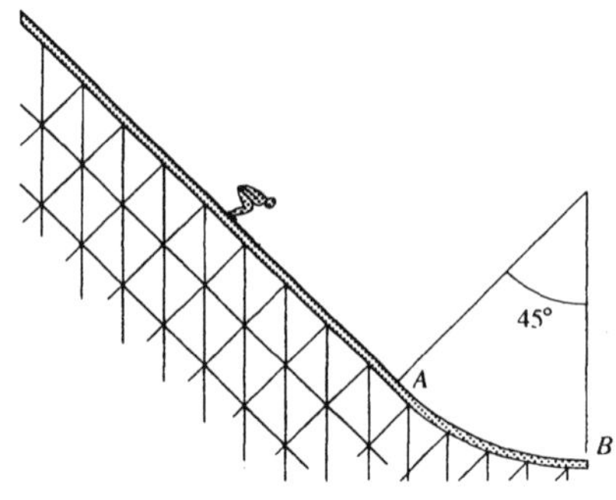

3.29ps The weights of the blocks are W_A and W_B and the surfaces are smooth. Let $R = W_A/W_B$. Draw a graph of the acceleration of block A as a function of R for $0.5 \leq R \leq 5$.

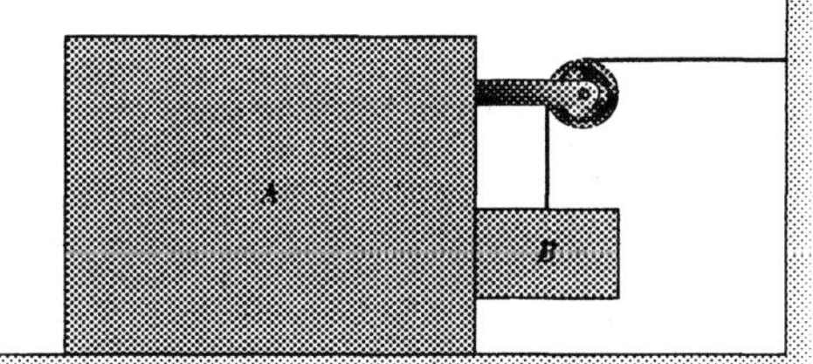

3.30ps For the roller coaster track shown, draw a graph of the number of g's experienced by the passengers at the top of the loop as a function of their speed v at the top of the loop for $30 \leq v \leq 120$ ft/s. (Define the number of g's to be the ratio of the downward normal force exerted on a passenger by the roller coaster to the passenger's weight. Negative g's result when the normal force is upward.)

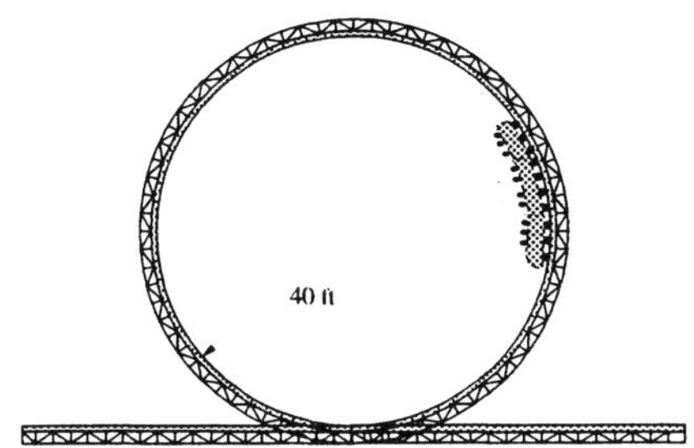

Chapter 4: Energy Methods

4.1ps Assume that the 1.19-million lb Big Boy locomotive is moving at 60 mi/hr and that the friction force exerted on the train's wheels during an emergency stop is 135,000 lb. Use work and energy to determine the distance the train requires to stop.

4.2ps Assume that the 1.19-million lb Big Boy locomotive shown in Problem 4.1ps is moving at 60 mi/hr when the brakes are applied, locking all its wheels. The coefficient of kinetic friction between the wheels and the track is 0.10. Use work and energy to determine the distance the train requires to stop.

4.3ps The 20,000 lb plane has a takeoff speed of 200 ft/s. The thrust of its engine is 24,000 lb. Neglecting aerodynamic drag, use work and energy to determine the length of runway required for the airplane to accelerate from rest to takeoff speed.

4.4ps Suppose that the aircraft in Problem 4.3ps is fitted with an auxiliary rocket engine that weighs 600 lb and provides a constant horizontal thrust T throughout the takeoff run. If the rocket decreases the required runway length by half, determine T. (Notice that you are neglecting the change in the mass of the airplane and the rocket engine.)

4.5ps The crate has a mass of 20 kg, the spring constant is 100 N/m, and the inclined surface is smooth. the crate is released from rest with the spring unstretched. Suppose that the device is first tested on Earth and then on Mars, where the acceleration due to gravity is 4.02 m/s^2. Determine the ratio of the distance the crate travels down the plane before stopping on Earth to the distance traveled on Mars. How is this ratio related to the ratio of the gravitational acceleration on Earth to that on Mars?

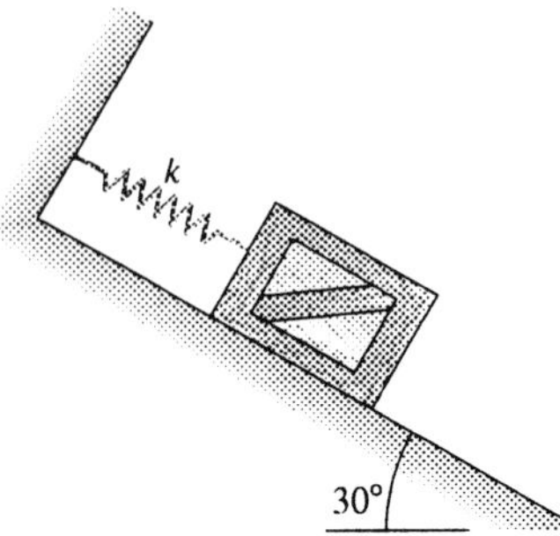

4.6ps For the apparatus described in Problem 4.5ps, determine the ratio of the maximum velocity attained by the crate on Earth to the maximum velocity attained on Mars. How is this ratio related to the ratio of the gravitational acceleration on Earth to that on Mars?

4.7ps In cases (a) and (b) the 30-kg box starts from rest at position 1 and the surfaces are smooth. For each case, use work and energy to determine the horizontal and vertical components of the velocity at position 2.

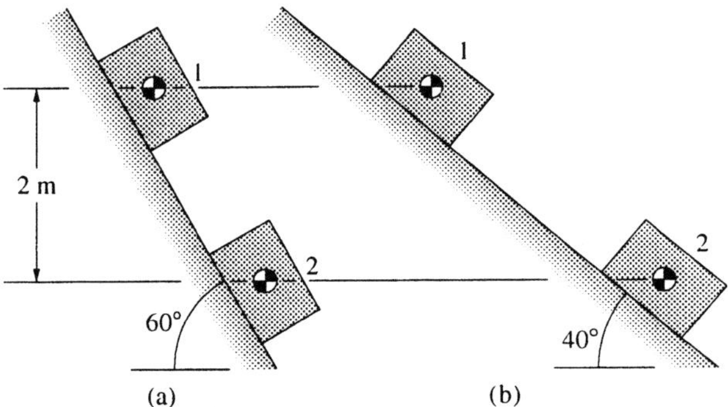

4.8ps The ball of mass m is released from rest in position 1. Assume that the length of the string is 2 meters. Draw a graph of the angular velocity of the string as a function of the angle α for $0 \leq \alpha \leq 90°$.

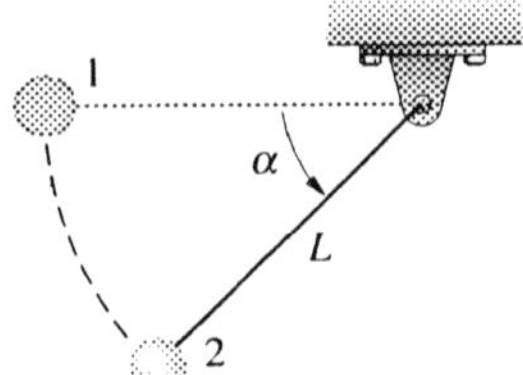

4.9ps The ball of mass m shown in Problem 4.7ps is released from rest in position 1 on the surface of the Moon, where the acceleration due to gravity is 5.31 ft/s^2. Draw graphs of the angular velocity of the string and the tension of the string as a function of the angle α for $0 \le \alpha \le 90°$.

4.10ps Starting from rest at position 1, the 200-kg wrecker's ball falls to position 2 and strikes a rock ledge. The cable is 6 meters in length. Determine: (a) the magnitude of the velocity v_E with which the ball would strike the wall on Earth; (b) the magnitude of the velocity v_M with which the ball would strike the wall on the Moon where the acceleration due to gravity is 1.62 m/s^2; (c) the ratio v_M/v_E

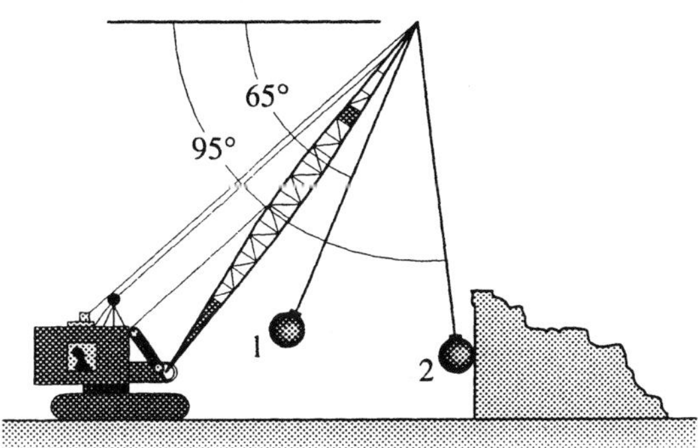

4.11ps The skier has a speed of 20 m/s at position 1. Neglecting friction and aerodynamic drag, determine: (a) how high (vertically) above position 1 she started down the incline; and (b) her speed when she reaches position 2.

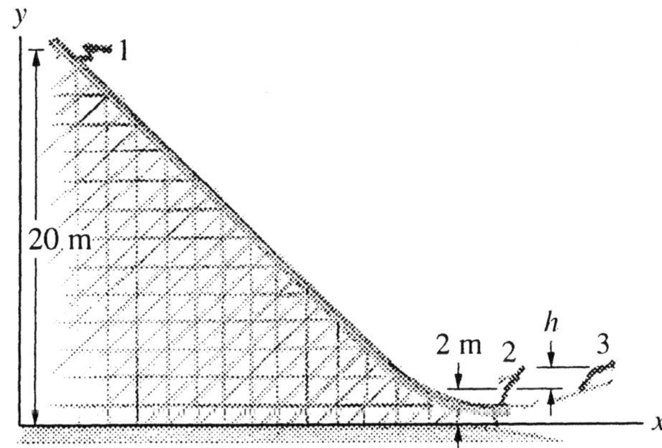

4.12ps A stunt driver wants you determine a velocity v_0 at which the car can enter the loop of radius R and coast through the loop without loosing contact with the track and without subjecting the driver to a normal acceleration greater than 1.5 g's. What range of velocities v_0 satisfies these requirements?

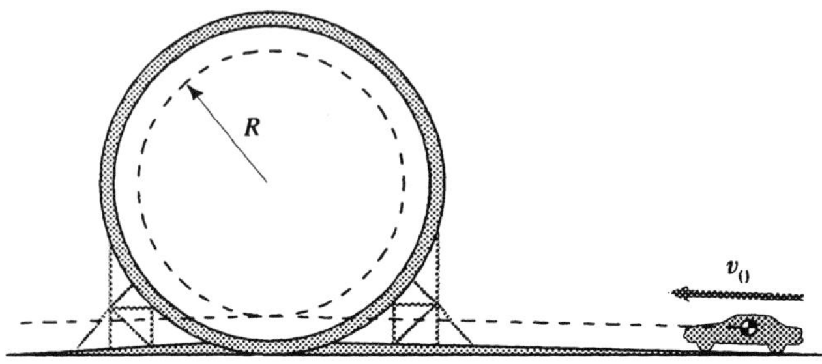

4.13ps The number of g's the driver in Problem 4.12ps is subjected to as the car enters the loop and at the top of the loop equals the normal forces exerted on him by the car divided by his weight. Suppose that he asks you to draw graphs of the number of g's he is subjected to as he enters the loop and at the top as a function of the entry speed v_0 in mi/hr. Let $R = 30$ ft.

4.14ps A small pellet of mass m is released from rest at position 1 and slides down the smooth surface of the cylinder. Compare the values of the angle α at which the pellet would leave the surface on Earth and on Mars, where the acceleration due to gravity is 4.02 m/s^2.

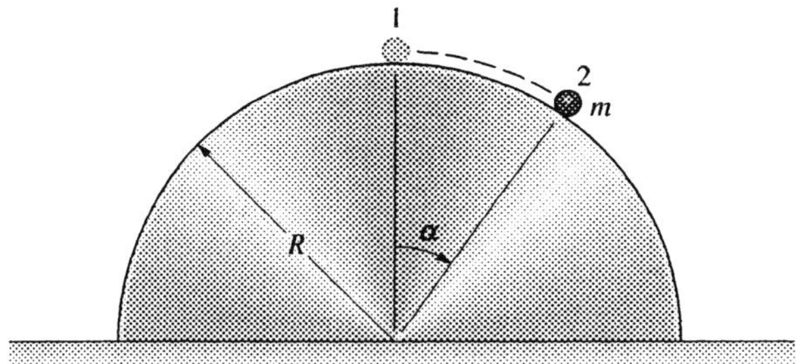

4.15ps Determine how high the spring-powered mortar launches the 10-lb package of fireworks above its initial position. The unstretched length of the spring is 30 inches and the spring constant is 1,500 lb/ft.

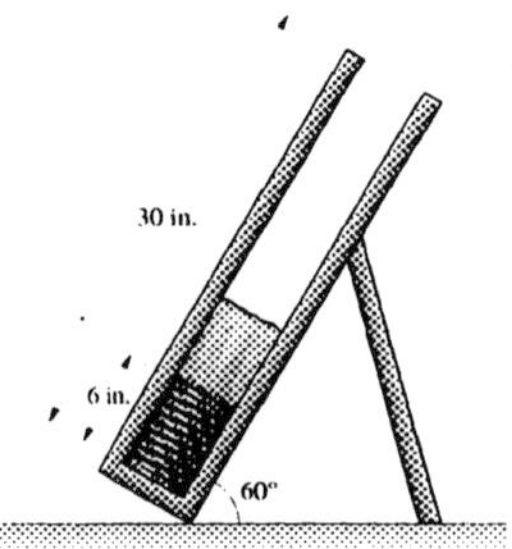

4.16ps If you moved the spring-powered mortar of Problem 4.13ps to the Moon where the acceleration due to gravity is 5.31 ft/s^2, how high would it launch the same 10-lb (Earth weight) package of fireworks above its initial position?

4.17ps Top-fuel dragsters can travel 1/4 mile from a standing start in less than 5 seconds. If a 1,000-lb dragster covers 1/4 mile in 4.96 seconds and you assume that its acceleration is constant, how much work is done on the car?

4.18ps In Problem 4.17ps, what average power (in horsepower) is transferred to the car as it covers 1/4 mile in 4.96 seconds?

4.19ps Assume that the engine is shut off when the Lunar Module has a downward velocity of 2 m/s and is a height h above the surface of the Moon (acceleration due to gravity = 1.62 m/s^2). Using conservation of energy, determine and draw a graph of the impact velocity as a function of h for $0 \leq h \leq 10$ meters.

4.20ps The potential energy associated with a force **F** acting on an object is $\mathbf{V} = 2x^2 - x$ N-m, where x and y are in meters. Suppose that the object moves in a straight line from position 1 to position 2 (along the diagonal of the square). Determine the work done by **F**.

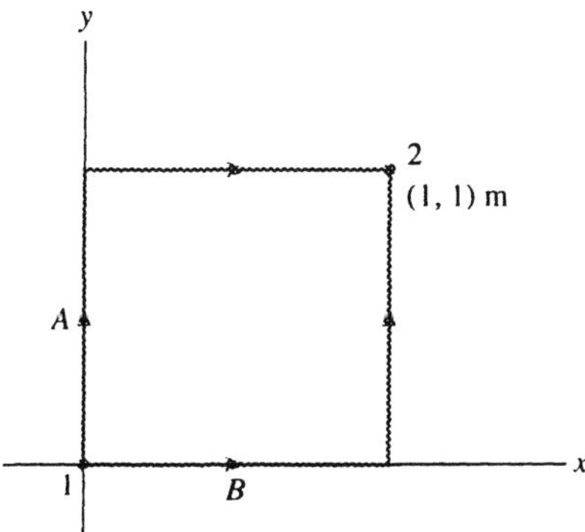

4.21ps Determine which of the following force fields (if any) are conservative.

(a) $\mathbf{F} = (5x^2y^3 - 2xy)\mathbf{i} + (5x^3y^2 - x^2)\mathbf{j}$

(b) $\mathbf{F} = (3x^4y^2z - 2xyz)\mathbf{i} + (3x^3y^2 - x^2)\mathbf{j} + (3x^3y^2 - x^2y)\mathbf{k}$

(c) $\mathbf{F} = -\cos(xy^2)\mathbf{i} + \sin(x^2y)\mathbf{j}$

4.22ps A student runs at 20 ft/s, grabs the rope, and swings out over the lake. Determine the angle θ at which he should release the rope to maximize the height he reaches. How high above his position when he grabs the rope, will he rise, and what is the resulting horizontal distance b?

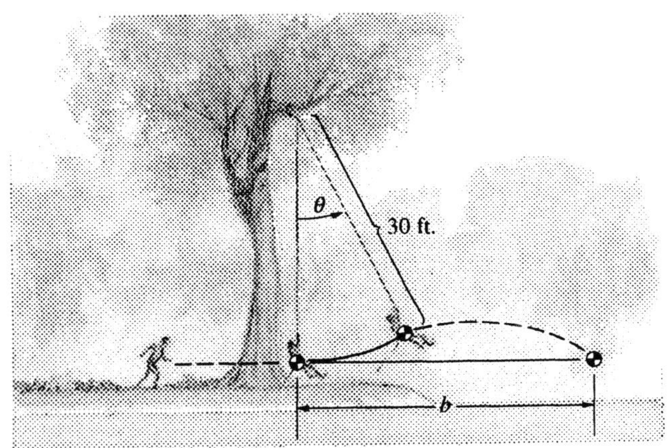

4.23ps The student in Problem 4.22ps runs at 20 ft/s, grabs the rope, and swings out over the lake. Draw a graph of the horizontal distance b as a function of the angle θ at which he releases the rope and use it to determine the value of θ which maximizes b. What is the maximum value of b?

4.24ps Suppose that the apparatus shown is tested on Earth and on the Moon where the acceleration due to gravity is 5.32 ft/s^2. The weight is 30 lb when weighed on the Earth. The spring constants are $k_A = 30$ lb/ft and $k_B = 15$ lb/ft. If the weight is released with the springs unstretched, determine the distance it falls before rebounding on Earth and on the Moon.

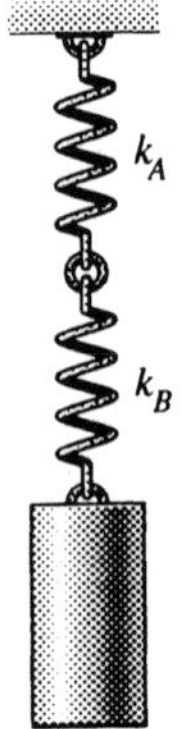

4.25ps Taking a running start at 1, the boy finds in his first ride that he reaches a horizontal distance $b = 25$ ft from position 2. He then changes the ramp angle from 35° to 40° (without changing the 5-ft height of the ramp) and in his second ride reaches a horizontal distance $b = 29$ ft. What was the change (if any) in his speed at 1 from his first ride to his second ride?

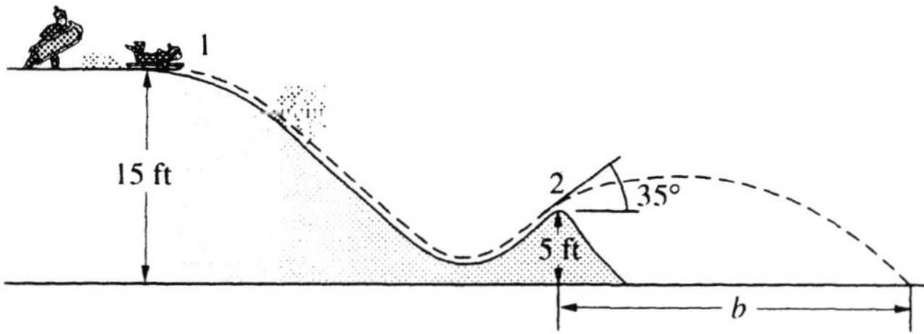

Chapter 5: Momentum Methods

5.1ps The dragster weighs 1200 lb. It can accelerate from rest to 301 mi/hr in 4.93 seconds, covering a distance of 1/4 mile.
(a) What impulse is applied to the dragster during its 4.96-second run?
(b) What is the average with respect to time of the dragster's acceleration?
(c) What is the average with respect to time of the force exerted on the dragster by the track?

5.2ps The constant force $\mathbf{F} = 85\mathbf{i} + 160\mathbf{j} - 420\mathbf{k}$ (N) is the only force on a 50-kg mass. At time $t = 0$ the velocity of its center of mass is $\mathbf{v} = 4\mathbf{i} + 35\mathbf{j} - 20,$ $\mathbf{k}$ (m/s). Calculate the impulse applied to the mass from $t = 0$ to $t = 5$ s and use it to determine the velocity of the mass at $t = 5$s.

5.3ps Draw graphs of the *x*, *y*, and *z* components of velocity of the 50-kg mass in Problem 5.2ps as functions of time for $0 \leq t \leq 5$ seconds.

5.4ps The 100-lb box starts from rest and is subjected to the force indicated in the graph. Use impulse and momentum to determine the velocity of the box at $t = 8$ seconds if $\mu_s = 0.35$ and $\mu_k = 0.30$.

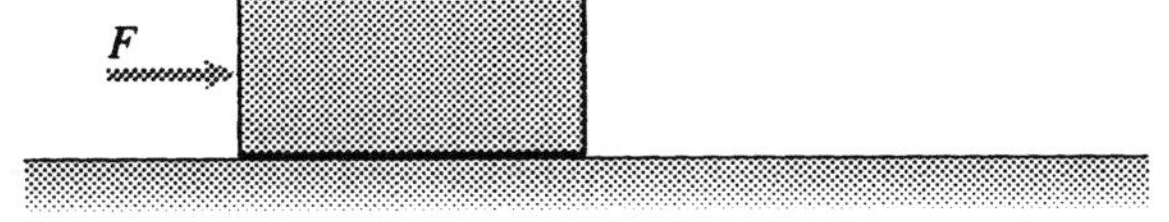

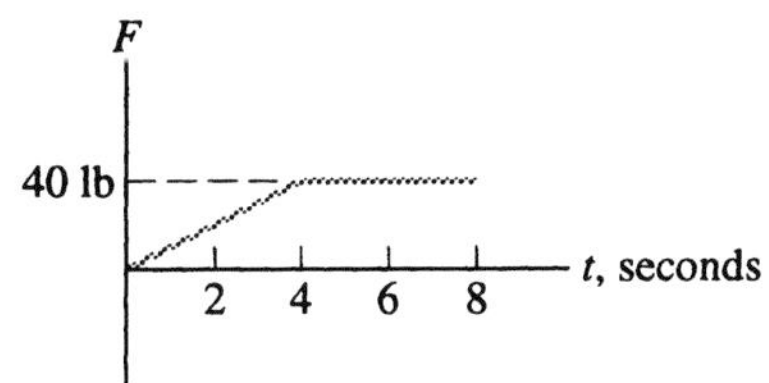

5.5ps Solve Problem 5.4ps with $\mu_s = 0.45$ and $\mu_k = 0.25$.

5.6ps The device stops 20-kg packages that start from rest and slide down the smooth ramp. A constant force is exerted at *B* to decelerate the package. If the maximum force that can be applied to a package without damaging it is 1,000 N, determine: (a) the minimum distance necessary to stop a package from the point of contact; and (b) the minimum time required to stop a package from the time of contact.

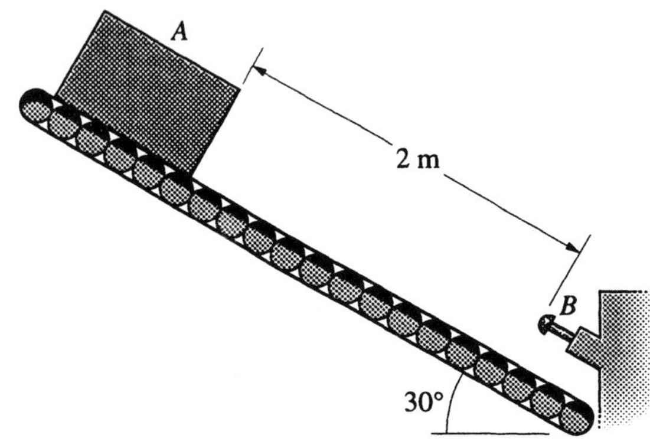

5.7ps In Problem 5.6ps, determine the linear impulse applied to a package in stopping it and also determine the deceleration necessary to stop it.

5.8ps A 30-gram projectile is launched by a rail gun. What linear impulse is applied to the projectile as it accelerates from rest to 5 km/s in 0.0005s?

5.9ps Assuming that the force exerted on the projectile in Problem 5.8ps is constant over the 0.0005 s interval, determine: (a) the acceleration of the projectile; (b) the distance traveled.

5.10ps The clock intended to measure the duration of the impact of the car with the barrier fails to work during a test. The data collected include the following: Car weight = 2,800 lb, impact speed = 5 mi/hr, and rebound speed = 1 mi/hr. Draw a graph of the average deceleration of the car during impact as a function of the assumed impact duration t_d for $0.1 \le t_d \le 0.5$ second.

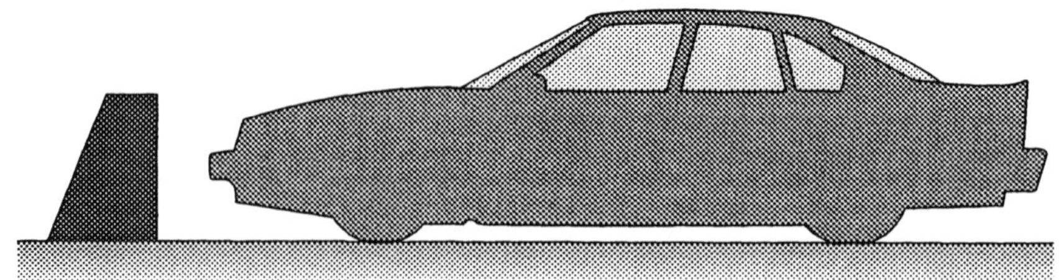

5.11ps The dummy and instrumented mask, designed to evaluate the suitability of the mask for hockey, are adapted by a bioengineer to measure facial impact forces due to the impact of a handball. A handball weighs 2.3 ounces, strikes the mask at 100 ft/s, and rebounds at 30 ft/s. The instrumentation measures the impact duration to be 0.05s. Determine the linear impulse exerted on the ball by the mask and the average force exerted on the mask by the ball during the impact.

5.12ps The ball used in the game Jai-alai is called a pelota. It has a 2-inch diameter, weighs 4.5 oz, and is caught and thrown with a cesta—a 2-ft long chestnut-framed curved wicker basket which forms an extension of the arm and enables the player to throw the ball at a very high speed. Assume that the pelota strikes the front wall at 250 ft/s and rebounds at 140 ft/s, with the impact lasting 0.04 s. What is the average force on the pelota during its impact with the wall? Also determine the linear impulse applied to the pelota by the wall.

5.13ps The 5-oz. baseball is struck 3 ft above home plate. The ball is moving horizontally at 148 ft/s when it is struck and the impact lasts 0.012 seconds. The ball leaves the bat at an angle of 30° above the horizontal. It clears the centerfield fence, 310 ft from home plate, at a height of 25 ft above the ground. Neglecting aerodynamic drag, determine the velocity with which the baseball left the bat and the magnitude of the average force exerted on the ball during its impact with the bat.

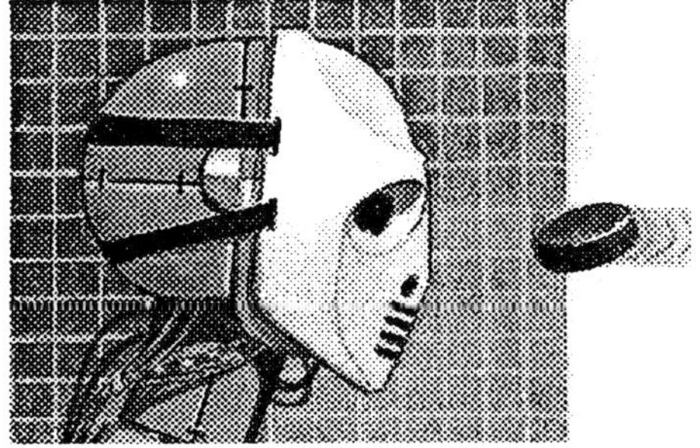

5.14ps In Problem 5.13ps, what is the average acceleration of the ball in g's during its impact with the bat?

5.15ps The cannon weighs 400 lb and the cannonball weighs 10 lb. The muzzle velocity of the cannonball is 200 ft/s. Neglecting the height of the muzzle above the ground and aerodynamic drag, draw a graph of the cannon's range as a function of the elevation angle (which is 10° in the figure) for elevation angles from 5° to 25° Also draw a graph of the cannon's horizontal recoil velocity as a function of the elevation angle.

5.16ps The bullet weighs 1 oz and the block weighs 10 lb. The block is initially at rest and the bullet is moving at 1000 ft/s. If $\mu_k = 0.30$ between the block and the surface on which it rests, determine the distance the block slides after the bullet hits it and becomes embedded.

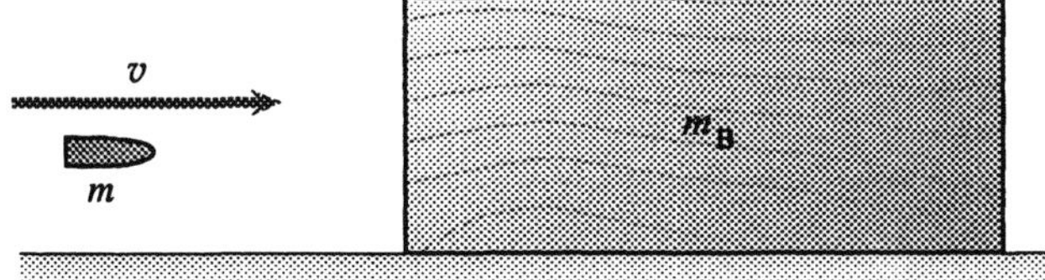

5.17ps In Problem 5.16ps, suppose that the bullet passes through the block and emerges with a speed of 100 ft/s. Determine the distance the block slides after being struck by the bullet. Neglect the speed list by the block while the bullet is inside the block.

5.18ps Suppose that you are a design engineer working to develop energy-absorbing bumpers. The two cars collide with $v_A = v_B = 5$ mi/hr. The weights of the cars are $W_A = 2{,}800$ lb and $W_B = 4{,}400$ lb. Draw graphs of the velocities of the cars after they collide as a function of the coefficient of restitution for $0 \le e \le 1$.

5.19ps The disk of mass m is attached to the string and slides on the smooth horizontal table. The string is drawn through a hole in the table at O at constant velocity v_0. Assume that the transverse force $F = 0$ and that at $t = 0$, $r = r_0$ and the transverse velocity of the disk (its velocity perpendicular to the string) is v_0. Determine the transverse velocity of the disk when $t = r_0/(2v_0)$.

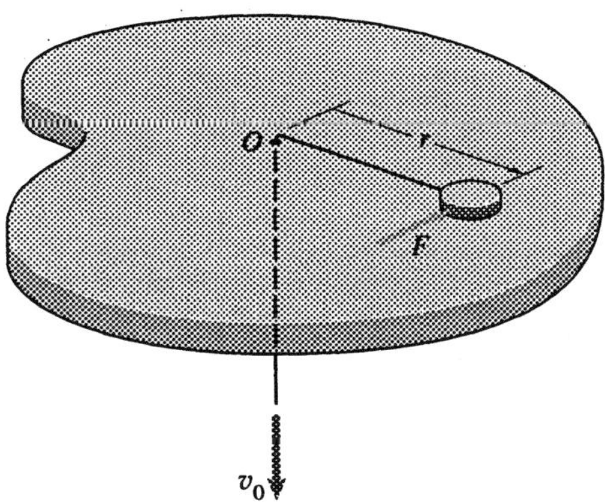

5.20ps An experimental Earth satellite has its perigee 300 km above the Earth's surface ($r_P = 6676$ km) and its apogee at twice the radius of the Moon's orbit (r_A = 768,000 km). Determine the velocities of the satellite at perigee and at apogee.

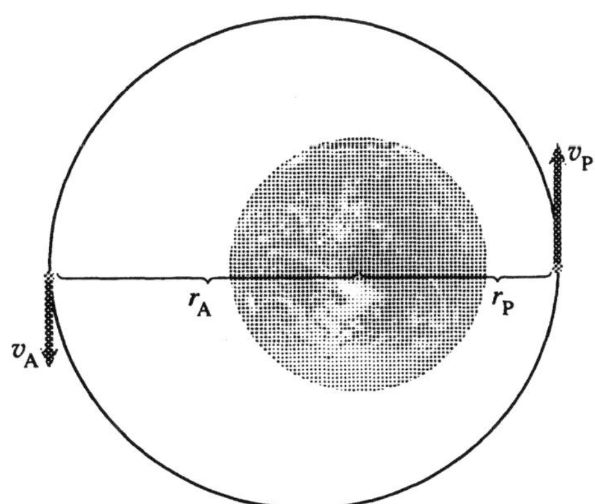

5.21ps The position of a 5-kg object is $\mathbf{r} = 7\mathbf{i} - 6\mathbf{j} + 3\mathbf{k}$ (m) and its velocity is $\mathbf{v} = 3\mathbf{i}\ |\ 6\mathbf{j}\ |\ 2\mathbf{k}$ (m/s). What is its angular momentum about the origin?

5.22ps What is the angular momentum of the 5-kg object in Problem 5.21ps about the point (6, 2, 3)?

5.23ps A 10-kg object is initially at rest at position $\mathbf{r} = \mathbf{i} + 2\mathbf{j} - 3\mathbf{k}$ (m). A force $\mathbf{F} = 3\mathbf{i} - 2t\mathbf{j} + t^2\mathbf{k}$ (N) is applied during the interval $0 \le t \le 5$ seconds.
(a) Use Newton's second law to determine the object's position and velocity

as functions of time over the interval $0 \le t \le 5$ seconds.
(b) By integrating $\mathbf{r} \times \mathbf{F}$ with respect to time, determine the angular impulse applied over the interval $0 \le t \le 5$ seconds.
(c) Use the results of part (a) to calculate the object's angular momentum about the origin at $t = 5$ s and compare it to the result of part (b).

5.24ps A turbojet engine in an airplane with an airspeed of 1,000 ft/s takes 1 slug of air per second into its intake. It burns 0.03 slugs of fuel per second and its exhaust velocity is 1,600 ft/s. What thrust is produced by the engine?

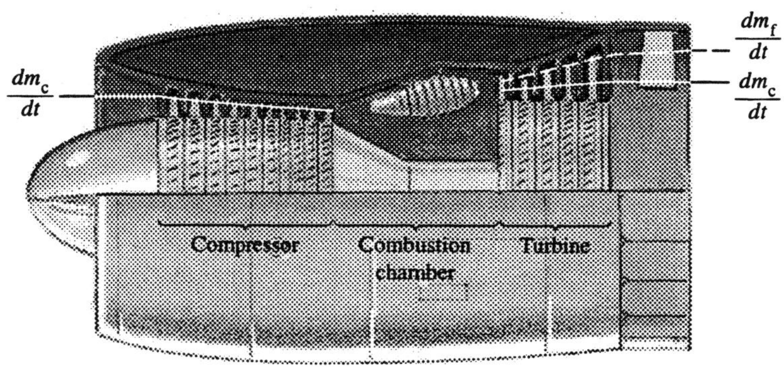

5.25ps The two-stage rocket has an exhaust velocity of 2,800 m/s. Assume that it starts from rest and neglect external forces. The mass of the payload is 2,000 kg, the mass of stage 1 is 33,000 kg, and the mass of stage 2 is 7,000 kg. Each stage is 90% propellant. Determine the velocity of the payload after all of the propellant has been expended.

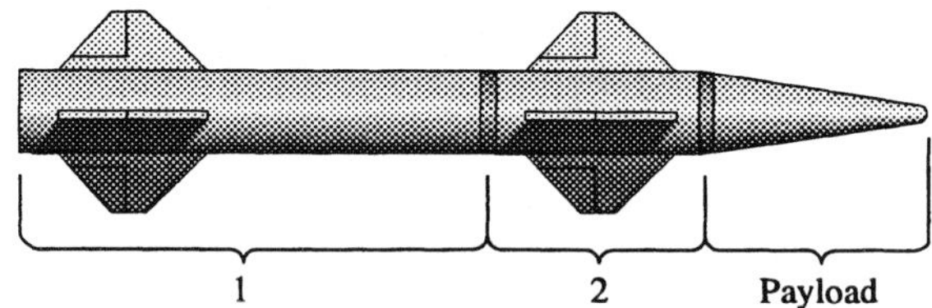

5.26ps Suppose that the rocket in Problem 5.25ps is replaced by a rocket that has a 2,800-m/s exhaust velocity, a 2,000-kg payload, and a single stage with a mass of 40,000 kg, 90% of which is propellant. Assume that the rocket starts from rest and neglect external forces. Determine the velocity of the payload after all of the propellant has been expended.

5.27ps The 47.5-Mg aircraft is moving at 70 m/s when it contacts an arresting system. The arresting system exerts an impulse of 2.85×10^6 N-m before the aircraft breaks through it. What is the aircraft's speed as the arresting system fails?

5.28ps A spacecraft is in an elliptic orbit around a planet. At the point closest to the planet, its radial distance $\mathbf{r}_P$ is 10,000 km. The maximum radial distance r_A reached by the spacecraft is 30,000 km. When the spacecraft is at rp, its velocity is 5 km/s. (a) What is the velocity of the spacecraft when it is at the maximum radial distance from the planet? (b) What is the transverse component of the satellite's velocity when it is at the radial distance $r = 20{,}000$ km?

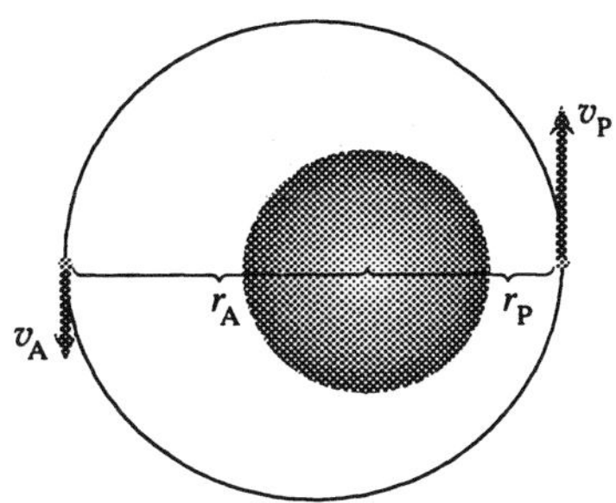

5.29ps In the shot-put event the athlete tries to throw the 16-lb shot as far as possible. Assume that the shot is released at a height of 7 ft above the stadium floor and that it strikes the ground 70 ft measured horizontally from the release point. Let the initial angle between the shot's path and the horizontal be θ_0. Draw a graph of the magnitude of the linear impulse needed to throw the shot 70 ft as a function of θ_0 for $35° \leq \theta_0 \leq 55°$.

5.30ps Assume that the athlete shown in Problem 5.29ps is in a future domed stadium on the surface of the Moon where the acceleration due to gravity is 5.31 ft/s^2. He releases the 16-lb (Earth weight) shot 7 ft above the stadium floor with velocity components $v_x = 31$ ft/s, $v_y = 26$ ft/s. (a) Determine the linear impulse applied to the shot as it is thrown. (b) Determine the horizontal distance from the release point to the point where the shot strikes the floor of the stadium.

5.31ps The water jet-powered ski boat draws water in at A with no horizontal velocity relative to the boat and expels it at B at 40 ft/s relative to the boat. The maximum mass flow rate through the engine is 2.5 slugs/s and the magnitude of the hydrodynamic drag force is $1.5v$ lb, where v is the boat's velocity in ft/s. The owner of the ski boat wants to increase the top speed of his boat. With his available budget he can do so either by modifying the engine to increase the flow rate to 3.0 slugs/s with the same maximum exhaust velocity, or by having the hull covered with drag-reducing paint which reduces the magnitude of the hydrodynamic drag force to $1.4v$ lb. Which option will produce the highest top speed?

5.32ps The boat owner in Problem 5.31ps decides to go into debt in order to modify the engine and paint the hull with drag-reducing paint. What is the top speed of the doubly modified boat?

Chapter 6: Planar Kinematics of Rigid Bodies

6.1ps Gear A starts from rest at $t = 0$ and has a clockwise angular acceleration given by $\alpha_A = Ct$ rad/s^2. Determine the constant C so that the hook will rise one meter in the first two seconds.

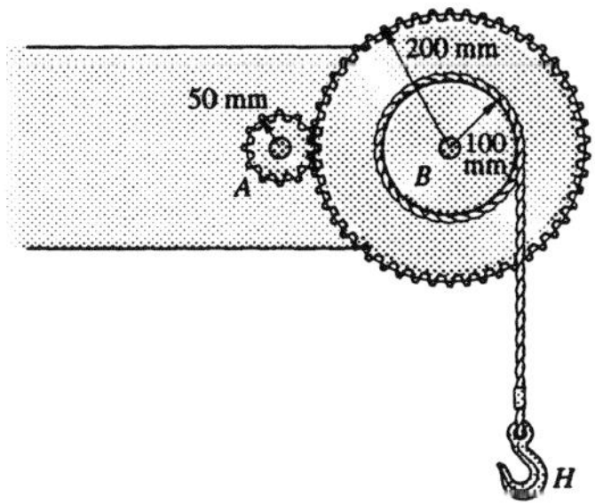

6.2ps The disk rotates about the fixed shaft O. At $t = 0$ it has a clockwise angular velocity of 3 rad/s. The angular acceleration of the disk is $0.5t$ rad/s^2 counterclockwise. Determine the magnitude of the acceleration of point A at $t = 7$ seconds.

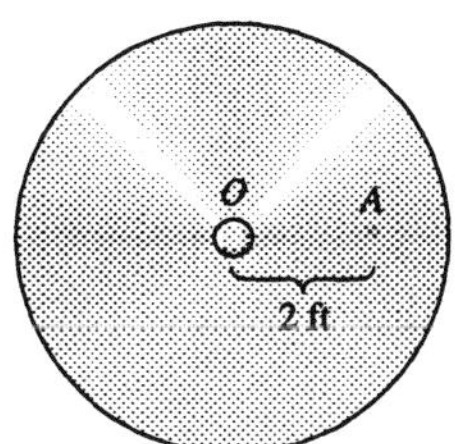

6.3ps In Problem 6.2ps, find the time during the interval $0 \le t \le 3$ s at which the magnitude of the total acceleration of point A is a minimum.

6.4ps The weight A starts from rest at $t = 0$ and moves downward with an acceleration of $2t$ m/s^2. What are the angular acceleration and angular velocity of the disk at $t = 2$ s?

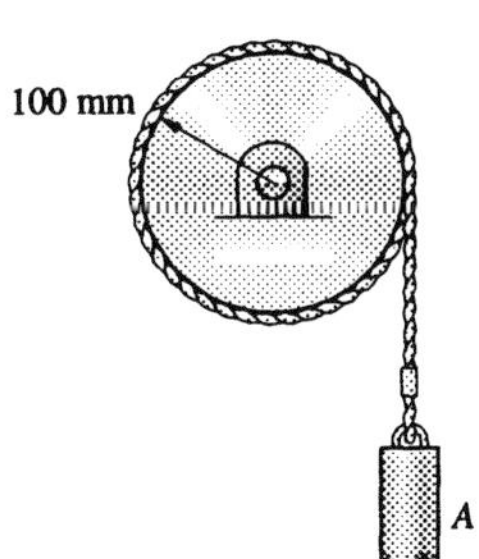

6.5ps The velocity $v_R = 3$ ft/s downward and the acceleration of the right-hand weight is 2 ft/s^2 downward. Determine the velocity and acceleration of the left-hand weight and also the angular velocities and angular accelerations of the two pulleys.

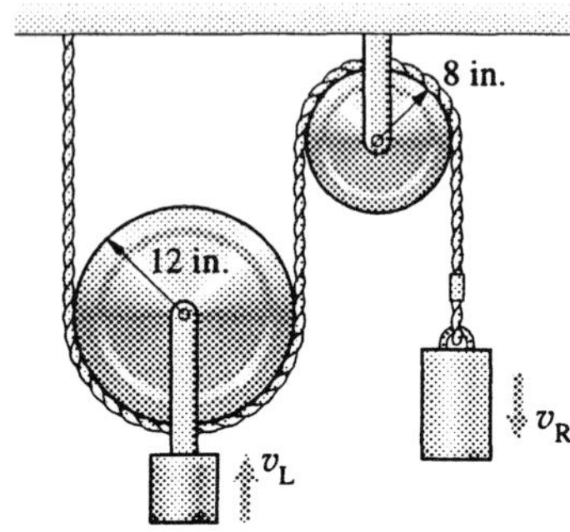

6.6ps Bar AB rotates with a constant clockwise angular velocity of 10 rad/s. Point C remains in contact with the surface. Let θ be the angle between bar AB and the line AC. Draw a graph of the velocity of point C as a function of θ for $0 < \theta \leq 120°$.

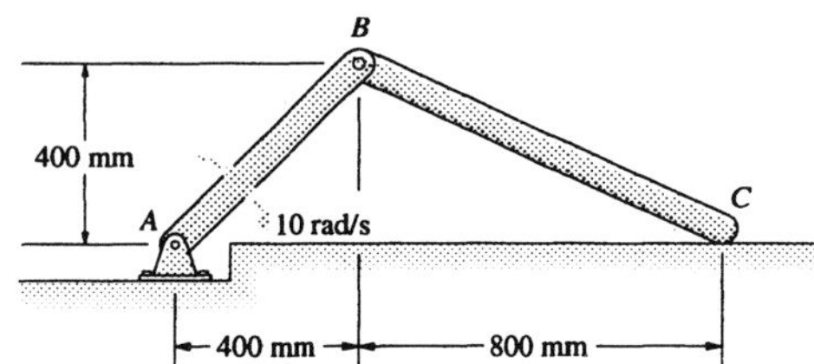

6.7ps Suppose that instead of the angular velocity of bar AB being 10 rad/s as shown, you know that bar AB is rotating in the clockwise direction and that $|\mathbf{v}_B| = 20$ ft/s. What is the velocity v_R of the rack?

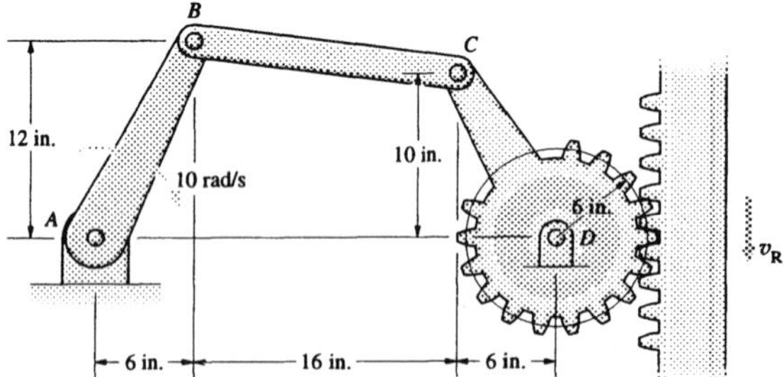

6.8ps The turbine rotates with an angular velocity of 30 rad/s about a fixed axis coincident with the x axis. (a) What is its angular velocity vector? (b) If the turbine was rotating in the direction opposite to the direction shown, what would be its angular velocity vector?

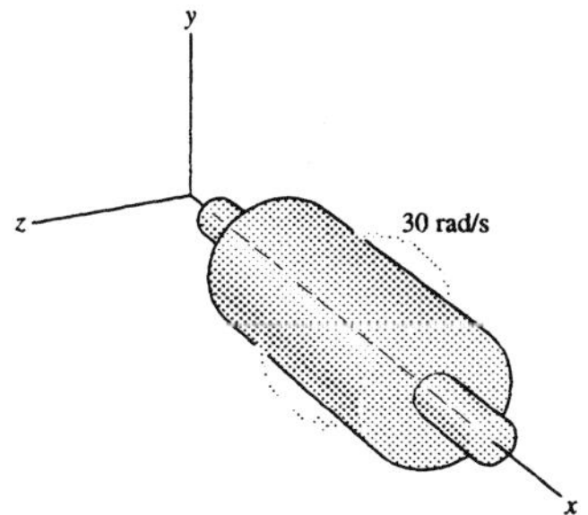

6.9ps The disk is rotating in the clockwise direction at 5 rad/s. At the instant shown, determine: (a) the velocity of point *B* relative to point *A*; (b) the velocity of point *C* relative to point *A*.

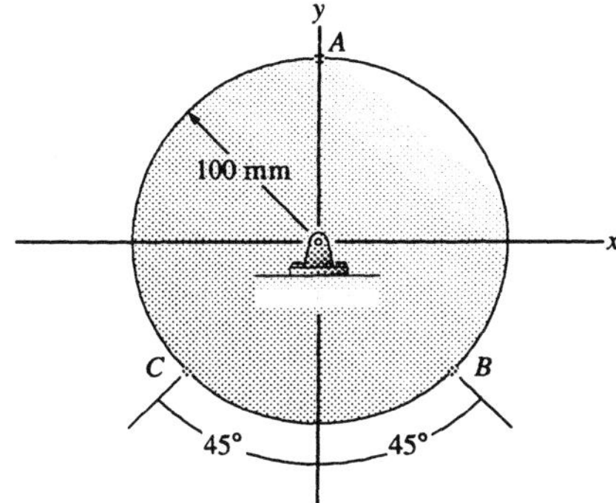

6.10ps At the instant shown, $v_A = 3$ ft/s to the right and the disk's angular velocity is $\omega = 6$ rad/s in the clockwise direction.

(a) Is the disk rolling?

(b) Determine the velocities of points *B*, *C*, and *D*.

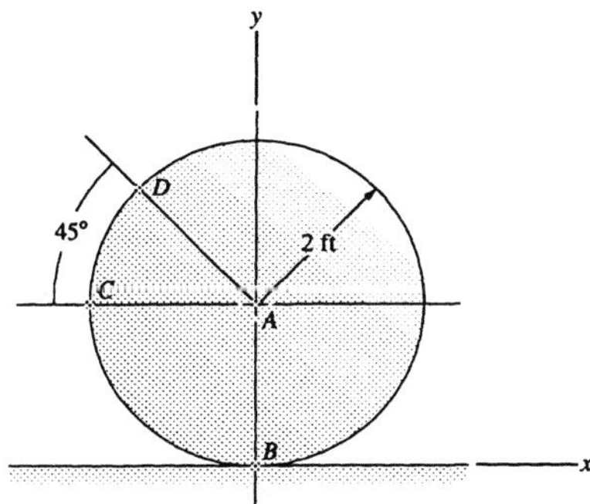

6.11ps Suppose that you don't know the angular velocity of bar *AB*, but you know that point *C* is moving at 2 in/s to the right. What is the angular velocity of bar *AB*?

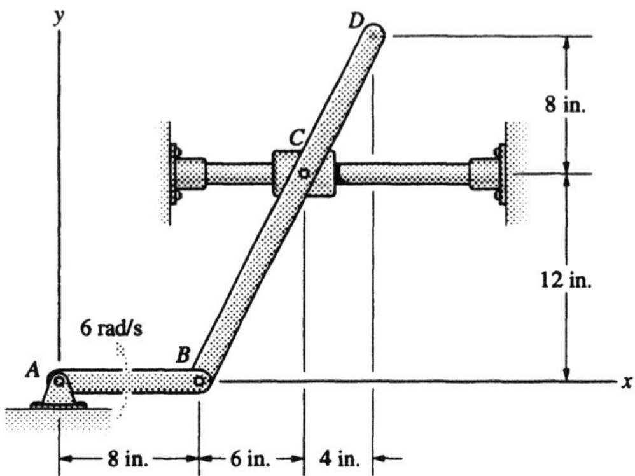

6.12ps Bar *AB* rotates at 10 rad/s in the counterclockwise. What is the velocity of the midpoint of bar *BC*?

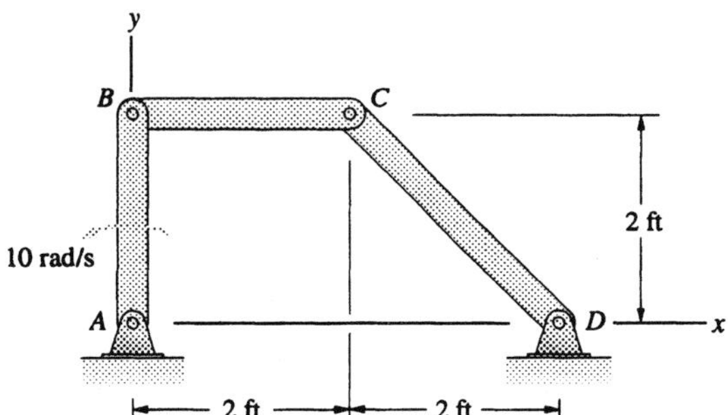

6.13ps Bar *AB* rotates at 12 rad/s in the clockwise direction. Determine the velocities of points *B*, *C*, and the midpoint of bar *BC*.

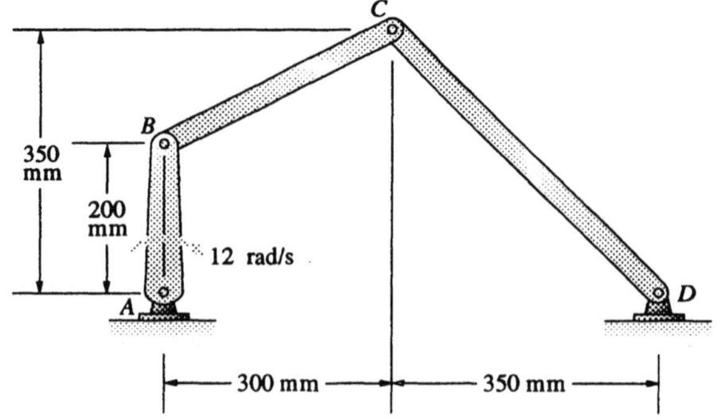

6.14ps In Problem 6.13ps bar AB rotates with a constant angular velocity of 12 rad/s. Determine the angular velocities of bars BC and CD when AB has rotated 90° clockwise from the position shown.

6.15ps Bar CD rotates at a constant clockwise angular velocity of 2 rad/s. As bar CD continues to rotate, the linkage will "jam" and the bars AB and BC will prevent bar CD from rotating further. Determine the angular velocities of bars AB and BC at the instant before jamming occurs.

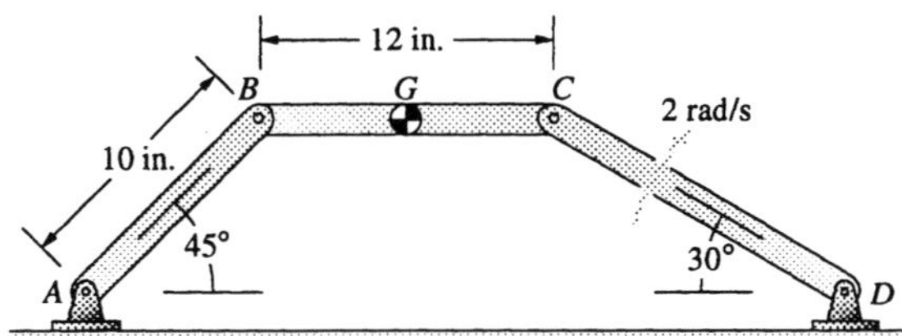

6.16ps Suppose that the constant 10 rad/s angular velocity of bar AB is clockwise instead of counterclockwise. Determine the angular velocity of bar BC and the velocity of point C when bar AB has rotated 90° from the position shown.

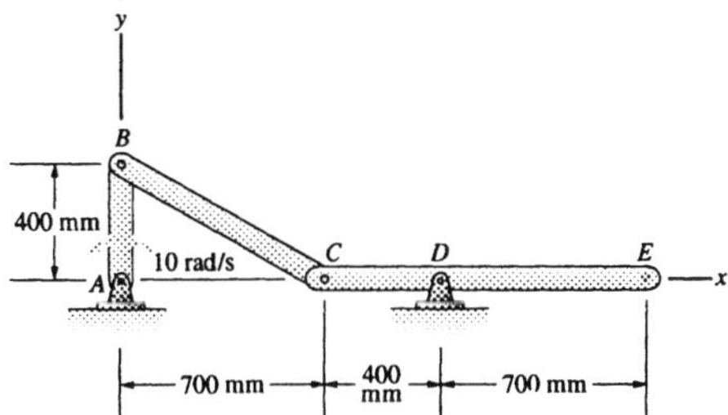

6.17ps In Problem 6.16ps, determine the angular velocity of bar CE and the velocity of point E when bar AB has rotated 90° from the position shown.

6.18ps Bar AB rotates in the counterclockwise direction at 2 rad/s. Determine the velocities of points B, C, and D.

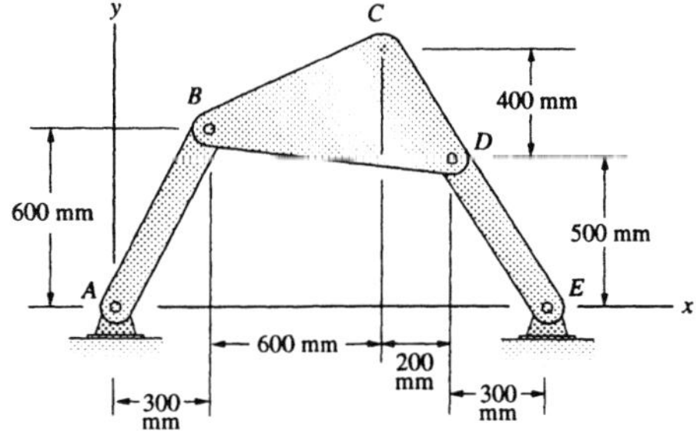

6.19ps If bar AB in Problem 6.18ps rotates with a constant counterclockwise angular velocity of 2 rad/s, the linkage will soon "jam" and bar AB cannot rotate further. Determine the velocity of point C at the instant before jamming occurs.

6.20ps The saw's motor rotates the disk mounted at A at a constant angular velocity of 10 rad/s in the counterclockwise direction. Determine the maximum resulting speed of point C to the left and to the right. (Link BC is 14 inches long and the radial distance AB is 4 inches.)

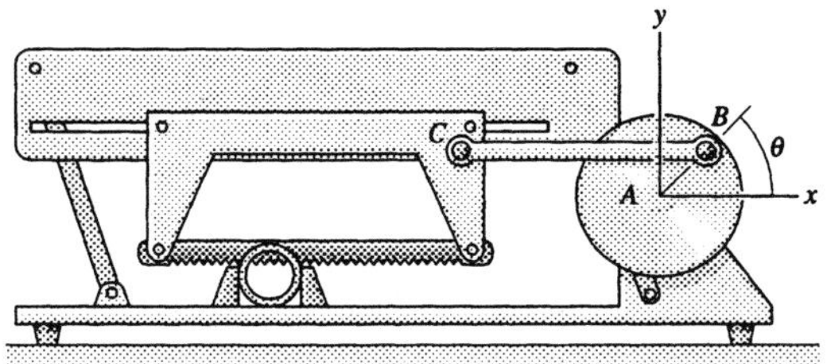

6.21ps The disk on the left rolls with a constant clockwise angular velocity of 2 rad/s. Draw a graph of the angular velocity of the right disk as the left disk rolls 360° from the position shown.

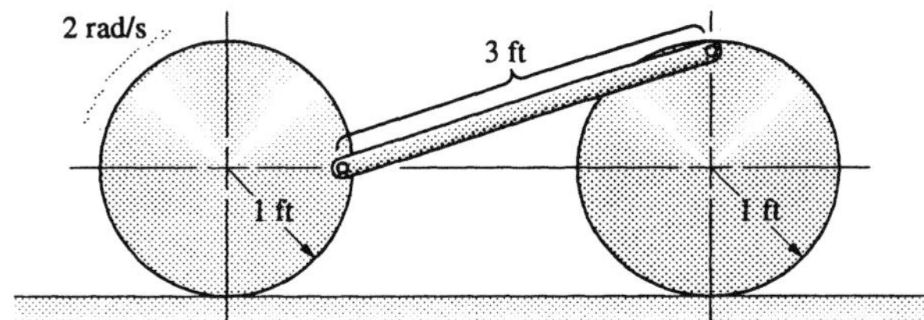

6.22ps Suppose that you don't know the angular velocity of bar AB, but you know that the angular velocity of bar BC is 5 rad/s in the counterclockwise direction. Use instantaneous centers to determine the angular velocities of bars AB and CD.

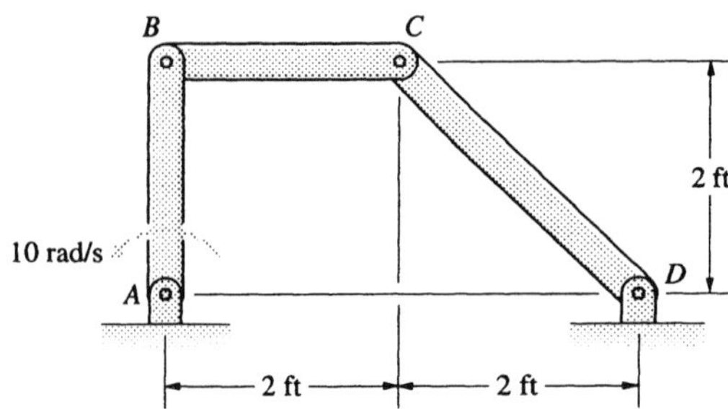

6.23ps The velocity of point A is $v_A = 120$ m/s. What is the angular velocity of the bar if the velocity of point B is $v_B = 20$ m/s? Determine the coordinates of the bar's instantaneous center.

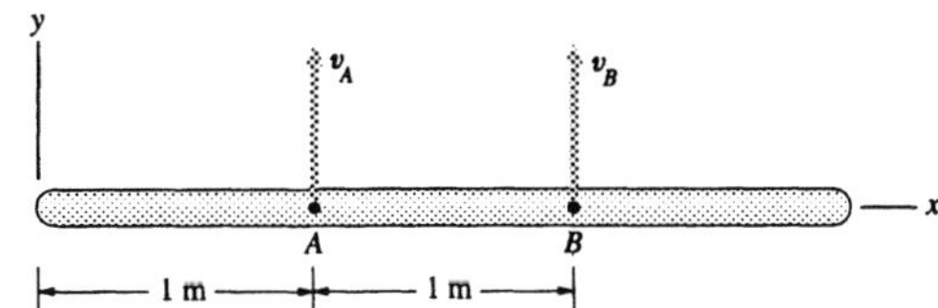

6.24ps Points A and B of the 2-m bar slide on the plane surfaces. Point A is moving to the left at 3 m/s.
(a) Determine the coordinates of the bar's instantaneous center.
(b) Use the instantaneous center to determine the velocity of point B.

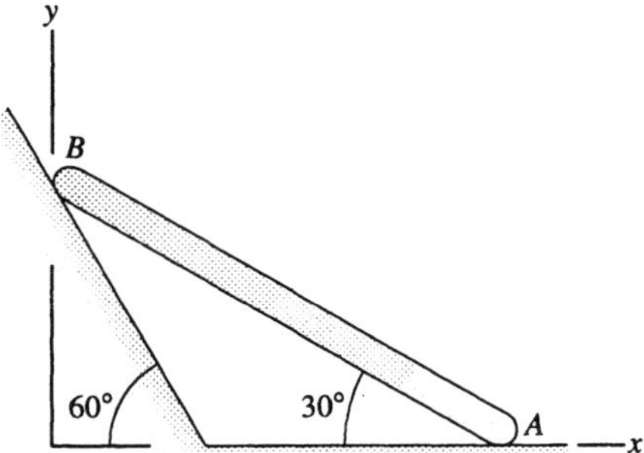

6.25ps For the lever arrangement shown, draw graphs of the x and y coordinates (relative to point O) of the instantaneous center of bar AB as a function of the angle AOB for $10° \le \theta \le 80°$.

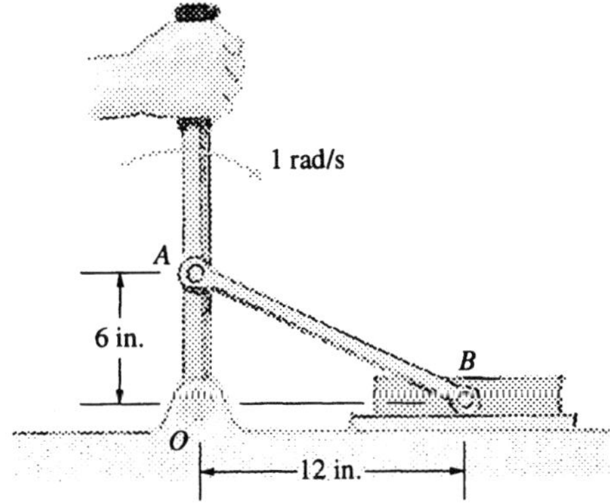

6.26ps Suppose that the slot in which the slider at C moves is vertical instead of horizontal, and C is at the same position as shown relative to A and B.

Bar *AB* is rotating at 6 rad/s in the clockwise direction. Use instantaneous centers to determine the angular velocity of bar *BC*.

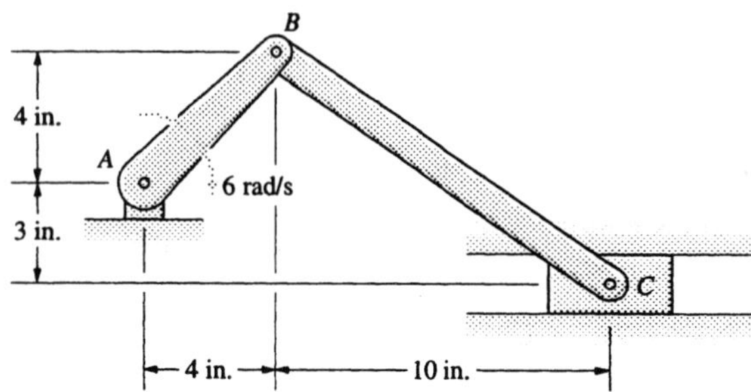

6.27ps The disk rolls on the level surface. Determine the velocity and acceleration of the point at the top of the wheel.

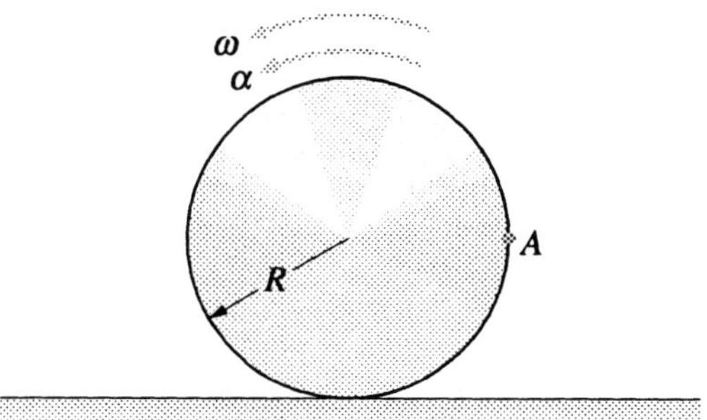

6.28ps The disk in Problem 6.27ps rolls with angular velocity $\omega = 2$ rad/s and angular acceleration $\alpha = 3$ rad/s^2. Let $r = 1$ m.
(a) At what point on the wheel is the magnitude of the velocity the greatest? What is the magnitude?
(b) At what points on the wheel is the magnitude of the acceleration the greatest? What is the magnitude?

6.29ps The disk in Problem 6.27ps rolls with angular velocity $\omega = 2$ rad/s and angular acceleration $\alpha = 3$ rad/s^2.
(a) At what point on the wheel is the magnitude of the velocity the least? What is the magnitude?
(b) At what points on the wheel is the magnitude of the acceleration the least? What is the magnitude?

6.30ps The left disk is rolling at a constant clockwise angular velocity of 2 rad/s. If the right disk rolls, what is the acceleration of its center at the instant shown? Refer to figure in problem 6.21ps.

6.31ps Bar *AB* has a counterclockwise angular velocity of 10 rad/s and a clockwise angular acceleration of 300 rad/s^2. Determine the velocity and acceleration of the midpoint of bar *BC*. (See Example 6.7, pages 264-265.)

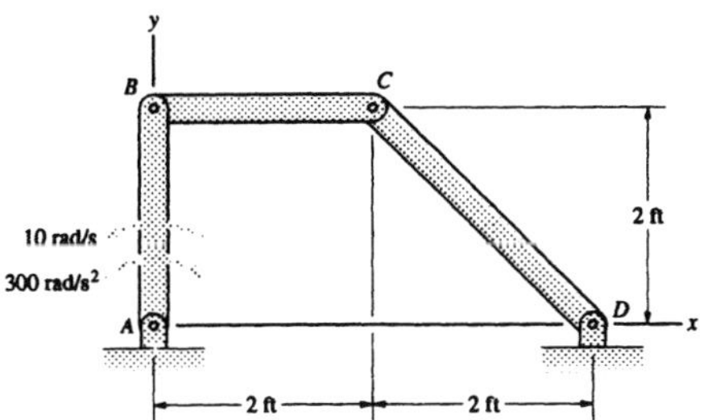

6.32ps The bar is rotating with a counterclockwise angular velocity of 20 rad/s. Draw a graph of the magnitude of the acceleration of point *B* as a function of the counterclockwise angular acceleration α of the bar for $-20 \leq \alpha \leq 20$ rad/s^2.

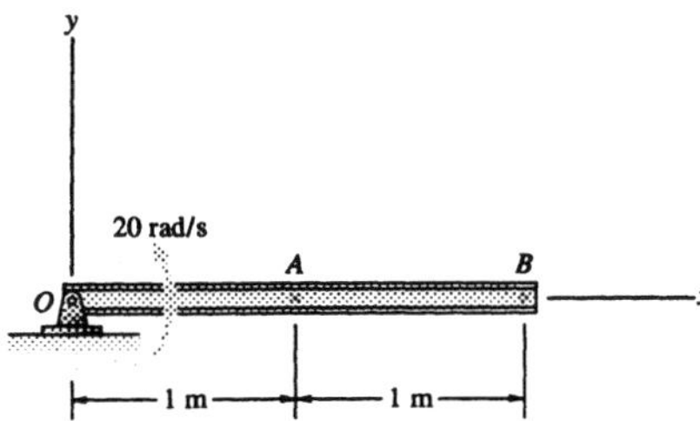

6.33ps The velocity and acceleration of point *A* are $\mathbf{v}_A = -3\mathbf{i}$ m/s and $\mathbf{a}_A = 2\mathbf{i}$ m/s^2. The angular velocity of the disk is 2 rad/s clockwise and its angular acceleration is 2 rad/s^2 counterclockwise.
(a) Is the disk rolling?
(b) Determine the accelerations of points *B*, *C*, and *D*.

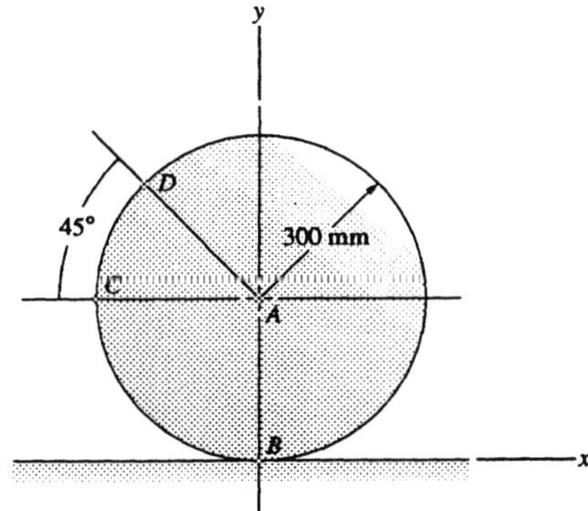

6.34ps Points A and B of the 2-m bar slide on the plane surfaces. At the instant shown, point A is moving to the left at 3 m/s and is accelerating to the right at 1 m/s². What are the velocity and acceleration of point B?

6.35ps The angular velocity and angular acceleration of link AB are $w_{AB} = 3$ rad/s and $\alpha_{AB} = 20$ rad/s². The dimensions of the plate are 15 inches by 25 inches. Determine the angular velocity and angular acceleration of link CD.

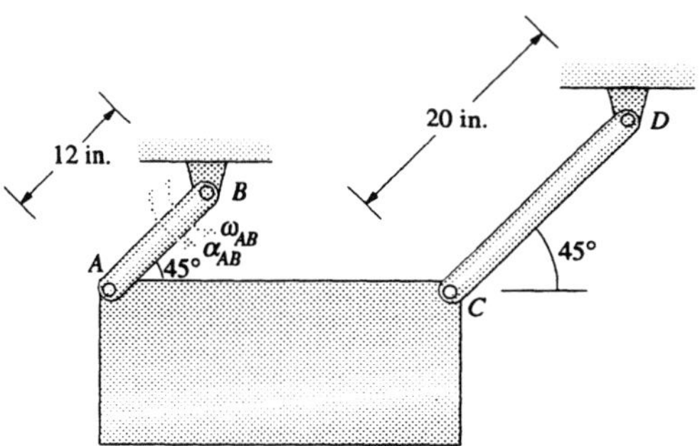

6.36ps In Problem 6.35ps, determine the velocity and acceleration of the lower right corner of the rectangular plate.

6.37ps Bar AB has a counterclockwise angular velocity of 10 rad/s and a clockwise angular acceleration of 20 rad/s². Determine the velocity and acceleration of the centerpoint of bar BC.

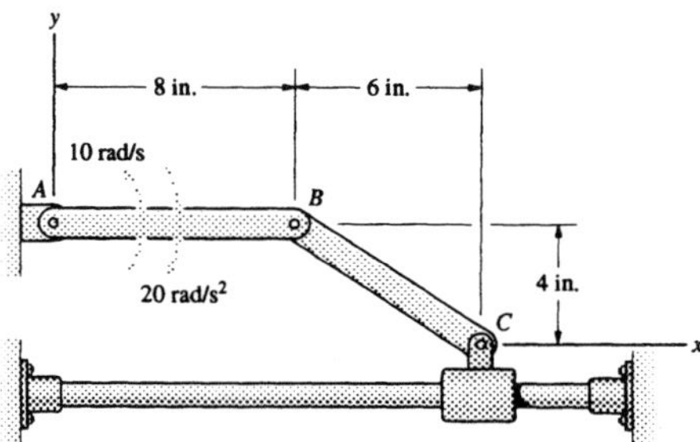

6.38ps Suppose that the angular velocity and angular acceleration of bar AB in Problem 6.37ps have the values shown and the system is in the position shown at $t = 0$. If the angular acceleration of bar AB is constant, determine the acceleration of point C at $t = 0.01$ s.

6.39ps Suppose that the angular velocity and angular acceleration of bar AB in Problem 6.37ps have the values shown and the angular acceleration of bar AB is constant. Determine whether the mechanism will "lock up," preventing further counterclockwise rotation of bar AB, before the angular acceleration results in stopping the counterclockwise motion of bar AB. If lock-up occurs, what is the angular velocity of bar AB the instant before it does?

6.40ps The slot in which the slider C moves extends sufficiently in both directions to allow crank AB to make full rotations. The system is in the position shown at $t = 0$ with $w_{AB} =$ rad/s and $\alpha_{AB} = 20$ rad/s^2, both clockwise, as shown. If α_{AB} is constant, what are the velocity and acceleration of point C at $t = 0.04$ s? Sketch the system at $t = 0.04$ s.

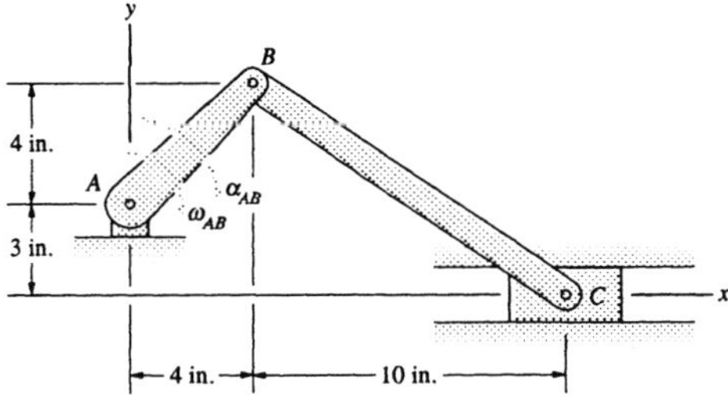

6.41ps In Problem 6.40ps, determine the maximum acceleration of the slider C during the interval $0 \le t \le 1$ s.

6.42ps In Problem 6.40ps, draw a graph of v_C as a function of time for $0 \le t \le 1$ s.

6.43ps The angular velocity and angular acceleration of the slotted bar AB at $t = 0$ are as shown. If the angular acceleration of bar AB is constant, what are the angular velocity and angular acceleration of bar AC at $t = 0.01$s?

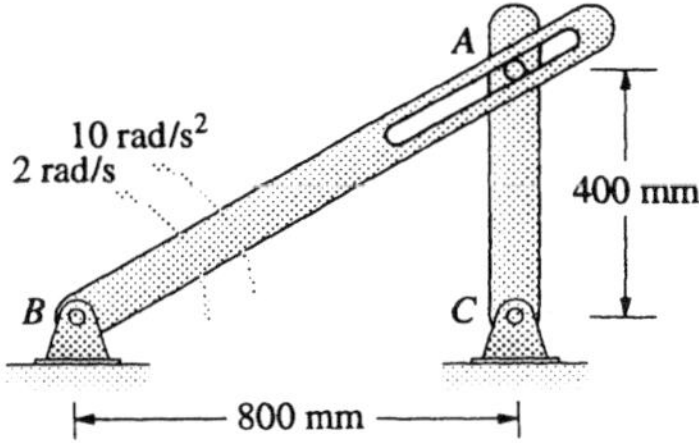

6.44ps The rod rotates with a constant counterclockwise angular velocity of 10 rad/s and the sleeve A moves outward along the rod at a constant rate of 4 ft/s. Draw a graph of the magnitude of the velocity vector of the sleeve A as a function of the distance AB for $1 \le AB \le 3$ ft.

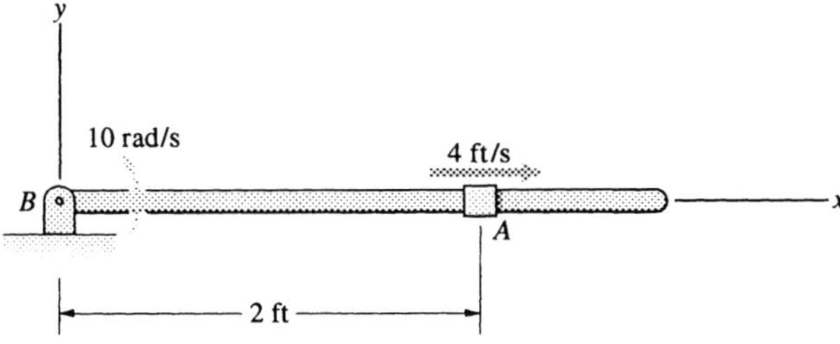

6.45ps In Problem 6.44ps, draw a graph of the magnitude of the acceleration vector of the sleeve *A* as a function of the distance *AB* for $1 \le AB \le 3$ ft.

6.46ps The sleeve *C* slides relative to bar *BD* at 1 m/s and bars *AB* and *BD* rotate at the angular velocities shown. Bar *BD* is 600 mm in length. Determine: (a) the velocity of *C* relative to point *B*; (b) the velocity of *C* relative to point *D*.

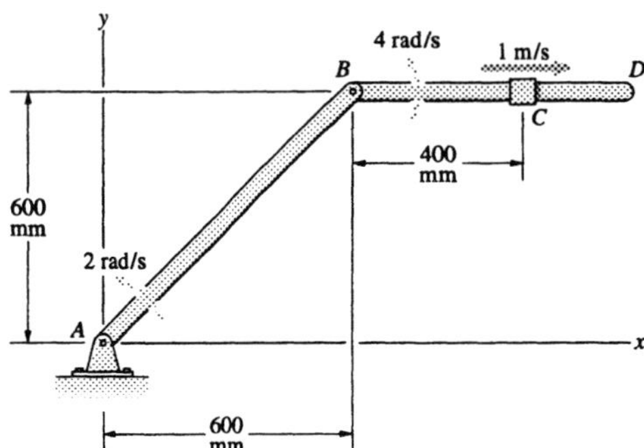

6.47ps In Problem 6.46ps, the velocity of the sleeve *C* relative to bar *BD* and the angular velocities of bars *AB* and *BD* are constant. At the instant shown, determine: (a) the acceleration of *C* relative to point *B*; (b) the acceleration of *C* relative to point *D*.

6.48ps The lineman's basket is moving straight upward at a constant speed of 0.25 m/s. At the instant shown, the coordinates relative to point *A* of the pivot holding the basket are $x = 3.6$ m, $y = 3.6$ m. What are w_{AC} and α_{AC}?

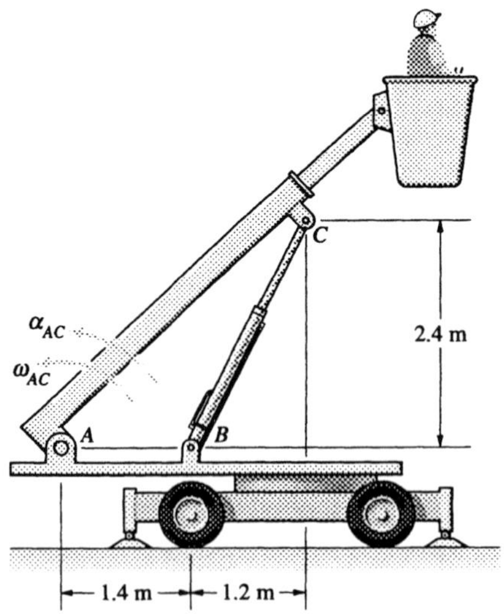

6.49ps In Problem 6.48ps, determine the rate at which the hydraulic actuator *BC* is extending.

6.50ps In Problem 6/48ps, determine the rate of change of the rate at which the hydraulic actuator *BC* is extending.

6.51ps Sleeve *A* moves upward at a constant speed of 10 m/s. Bar *AC* slides through the sleeve at *B*. Let the angle between bar *AC* and the horizontal (which is 30° in the figure) be denoted by θ. Draw a graph of the angular velocity of bar *AC* as a function of θ for $0 \leq \theta \leq 40°$.

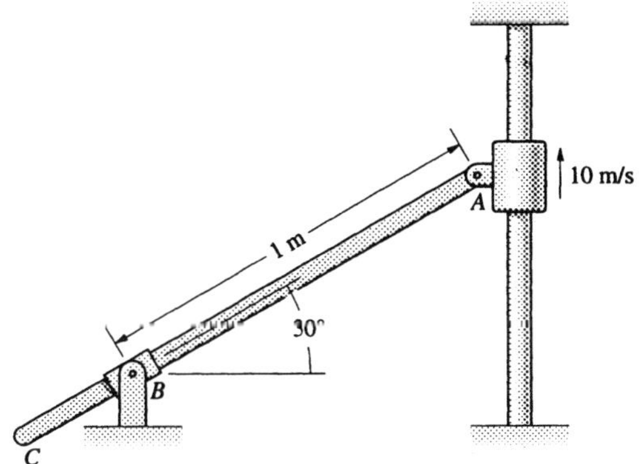

6.52ps In Problem 6.51ps, draw a graph of the magnitude of the velocity of bar *AC* relative to the sleeve *B* as a function of θ for $0 \leq \theta \leq 40°$.

6.53ps Block *A* slides down the incline at 2 ft/s and has an acceleration of 1 ft/s^2 up the incline at the instant shown. What is the velocity of point *C*?

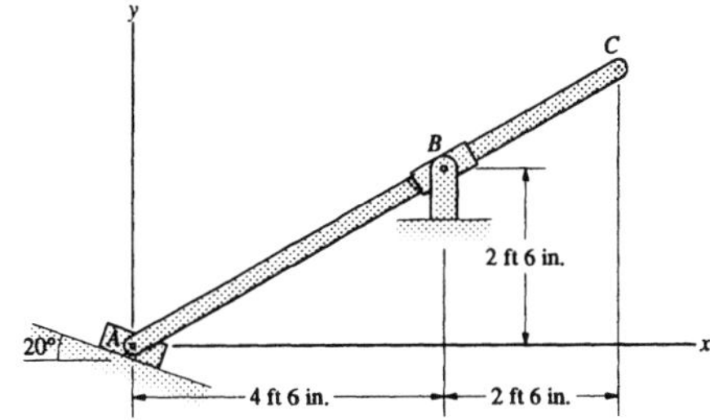

6.54ps In Problem 6.53ps, determine the velocity of the point of bar *AC* that is coincident with the pin *B* at the instant shown.

6.55ps In Problem 6.53ps, determine the acceleration of point *C* and the acceleration of the point of bar *AC* that is coincident with the pin *B* at the instant shown.

6.56ps The radius of the merry-go-round is $R = 5$ ft. At the instant shown, the merry-go-round is rotating at an angular velocity of 3 rad/s and its rate of rotation is slowing at 0.2 rad/s^2. In Case 1, the girl at *B* observes the motion of the boy standing on the ground just beside the merry-go-round at *A* using a coordinate system that rotates with the merry-go-round. What are the boy's velocity $\mathbf{v}_{A\text{rel}}$ and acceleration $\mathbf{a}_{A\text{rel}}$ relative to her rotating coordinate system?

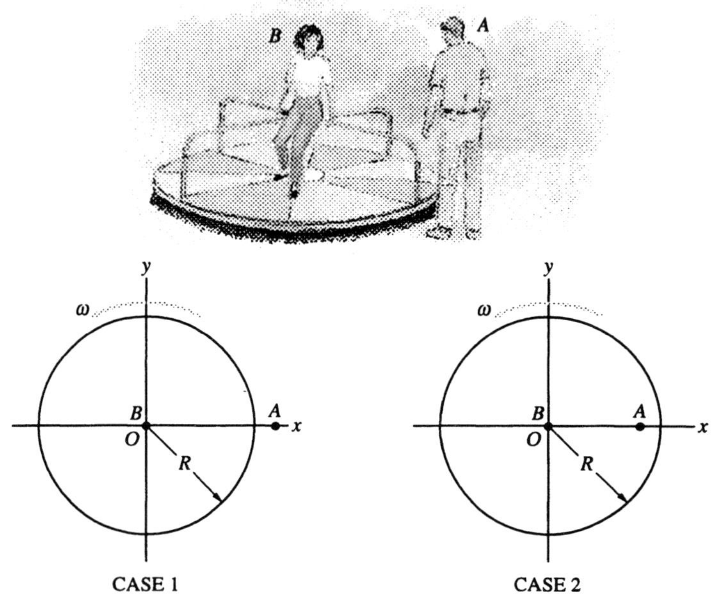

6.57ps In Case 2 of Problem 6.56ps, the girl at B observes the motion of the boy riding on the edge of the merry-go-round at A using a coordinate system that rotates with the merry-go-round. Use eqs (6.19) and (6.20) to determine the boy's velocity and acceleration relative to the Earth.

6.58ps At the instant shown, the ship B is moving west at a speed of 18 m/s and has a tangential component of acceleration of 1 m/s². The ship is turning toward the south at a constant rate of 2° per second. The helicopter A is hovering motionless (relative to the Earth). The helicopter's position relative to the ship-fixed coordinate system is $\mathbf{r}_{A/B} = 500\mathbf{i} + 200\mathbf{j} + 450\mathbf{k}$ (m). Determine the helicopter's velocity and acceleration relative to the ship-fixed coordinate system.

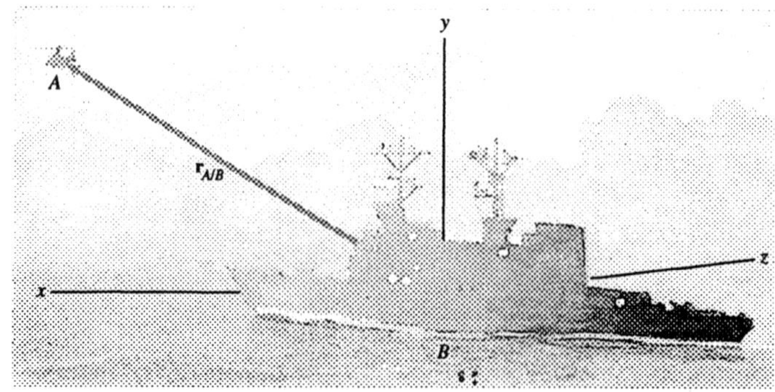

6.59ps In Problem 6.58ps, suppose that instead of being motionless relative to the Earth, the helicopter A is moving at 20 m/s straight toward the origin of the ship's coordinate system at the instant shown. Determine $\mathbf{v}_A$, $\mathbf{v}_{A/B}$, $\mathbf{a}_A$, and $\mathbf{a}_{A/B}$ at this instant.

6.60ps You are playing tennis on the deck of the cruise ship while the ship is docked at a port at the equator. The x axis of the coordinate system points north. Assume the Earth's angular velocity to be 2π radians per 24 hours. At the present instant the ball is located at $\mathbf{r}_{A/B} = 10\mathbf{i} + 5\mathbf{j} + 20\mathbf{k}$ (ft), and the velocity of the ball relative to the ship is $\mathbf{v}_{A\text{rel}} = 10\mathbf{i} - 0.5\mathbf{j} + 15\mathbf{k}$ ft/s. The ball weighs 0.125 lb and the aerodynamic force acting on it is $\mathbf{F} = -0.01\mathbf{i} + 0.005\mathbf{j} - 0.015\mathbf{k}$ (lb). Assuming that a parallel non-rotating coordinate system with its origin at the center of the Earth is inertial, determine the ball's acceleration: (a) relative to the non-rotating coordinate system with its origin at the center of the Earth; (b) relative to the coordinate system shown. (The Earth's radius is 3,960 miles.)

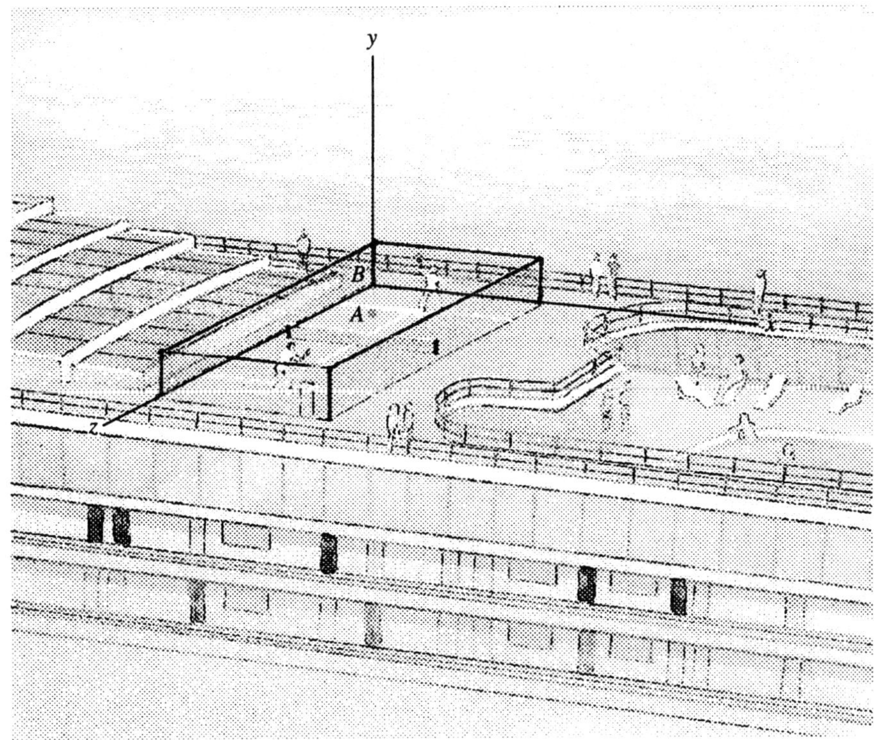

6.61ps The merry-go-round rotates with a constant counterclockwise angular velocity of 0.5 rad/s. A girl A walks out along a radial line at 1 m/s as shown. By using a coordinate system that rotates with the merry-go-round, determine the girl's acceleration relative to the Earth and drag a graph of its magnitude as a function of the radial distance AB for $1 \le AB \le 4$ m. If the coefficient of static friction between the girl's shoes and the floor of the merry-go-round is 0.6, how far out from the center can she walk without slipping? How far out can she stand without slipping? If there is a difference in the two distances, explain it.

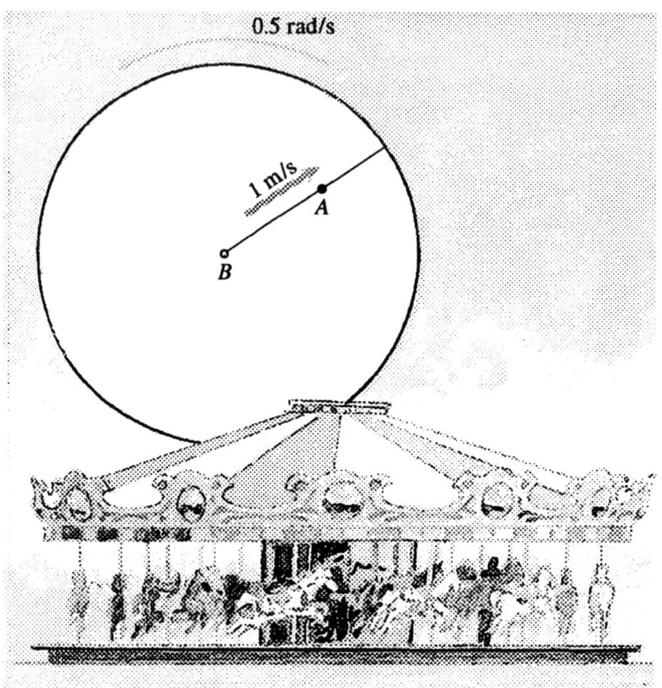

6.62ps Suppose that the merry-go-round of Problem 6.61ps has a radius of 6 meters and it is desired that people be able to walk at 1 m/s without slipping anywhere on the rotating merry-go-round. Assume the minimum static coefficient of friction between peoples' shoes and the floor of the merry-go-round to be 0.5. It is also desired that the magnitude of the velocity of the outer edge of the merry-go-round relative to the ground be no more than 2.5 m/s for the safety of those who unwisely step off the moving merry-go-round. Determine the maximum angular velocity at which the merry-go-round can operate.

6.63ps The disk-shaped space station of radius $R = 6$ meters rotates with a constant counterclockwise angular velocity $\omega = 0.25$ revolutions/second about the axis perpendicular to the page. At the instant shown, the person at A has just started moving at a constant speed of 1 m/s toward the center O of the station along the x axis of the body-fixed coordinate system shown. Using Eqs. (6.19) and (6.20) and the body-fixed coordinate system, determine A's velocity and acceleration relative to a non-rotating reference frame with its origin at O. Assume that the non-rotating reference at O is inertial.

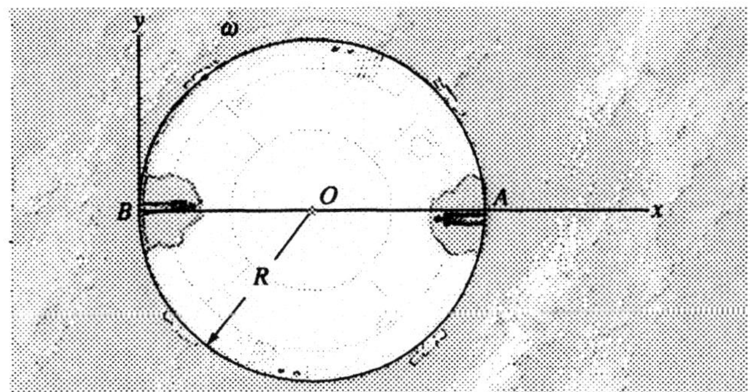

6.64ps Consider Problem 6.63ps. In terms of a non-rotating reference frame with its origin at *O*, what are *A*'s velocity and acceleration relative to the person *B*, who is not moving relative to the station?

6.65ps The ship *B* is traveling north at a constant velocity of 20 ft/s. At the instant shown, the aircraft is flying toward the east at a constant speed of 780 ft/s and is turning toward the north at a constant rate of 1.5° per second. The position of the ship relative to the aircraft is $\mathbf{r}_{B/A} = -3{,}000\mathbf{i} - 3{,}000\mathbf{j} - 35{,}000\mathbf{k}$ (ft). What is the velocity of the ship relative to a coordinate system that is fixed to the airplane with the the positive *x* axis pointing out the aircraft's nose, the positive *y* axis pointing north, and the positive *z* axis pointing upward?

6.66ps In Problem 6.65ps, what is the acceleration of the ship relative to the airplane fixed coordinate system?

6.67ps The space shuttle is deploying the satellite shown. Let the satellite be identified by *P* and the shuttle by *S*. At the instant shown, the satellite's position relative to the shuttle-fixed coordinate system is $\mathbf{r}_{P/S} = 50\mathbf{i}$ (m). The inertial measurement unit on the shuttle indicates that the shuttle's angular velocity is $\mathbf{w} = 0.0015\mathbf{i} - 0.0030\mathbf{j} + 0.0025$ (rad/s). The rendezvous radar on the shuttle determines that the velocity of the satellite relative to the shuttle-fixed coordinate system is $\mathbf{v}_{P/S} = -0.1\mathbf{i} - 0.15\mathbf{k}$ (m/s). Determine the x, y, and z components of the satellite's velocity relative to a nonrotating coordinate system with its origin at the shuttle's center of mass.

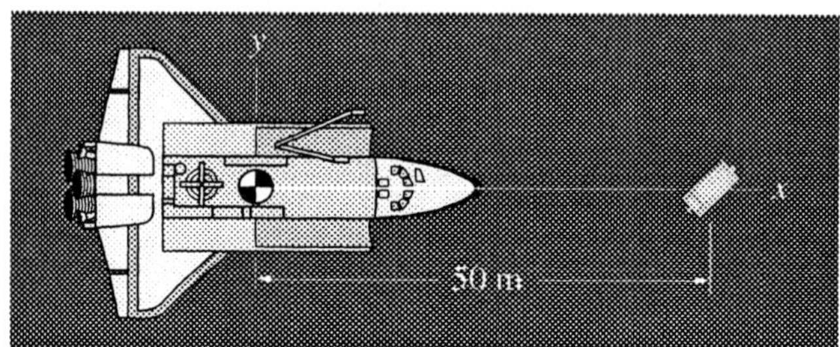

6.68ps In Problem 6.67ps, if the acceleration of the satellite relative to a nonrotating coordinate system with its origin at the shuttle's center of mass is zero, what is the satellite's acceleration relative to the shuttle-fixed coordinate system?

6.69ps The rotating laboratory has a constant angular velocity w of one revolution every six seconds. A man places an object on air bearings (friction is negligible) at A with an initial velocity (relative to the Earth) of $-2\mathbf{i}$ (m/s). Draw a graph of the object's path (until it strikes the outer wall; the radius of the laboratory is 6 m) relative to the coordinate system shown if: (a) the coordinate system does not rotate; (b) the coordinate system rotates with the laboratory.

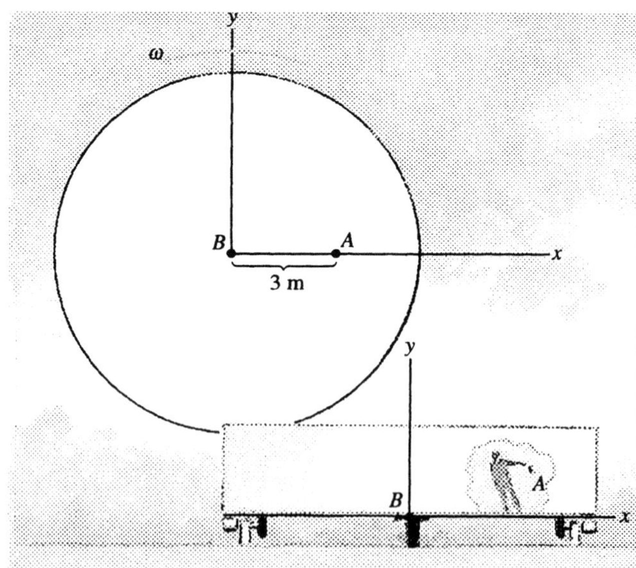

6.70ps Consider Problem 6.69ps. Determine two distinct initial velocities relative to the laboratory which the man can give the object that cause it to return to him.

Strategy: Consider various straight-line paths traveled by the puck relative to non-rotating coordinates and where these paths reach a distance of 3 meters from the center of the laboratory. Then determine where the man is as a function of time relative to non-rotating coordinates.

6.71ps Using the pulley system shown as a model, design and sketch a new pulley system in which the velocity of the hook is $v_H = 20$ mm/s when the cord coming around the large pulley is pulled downward at 120 mm/s.

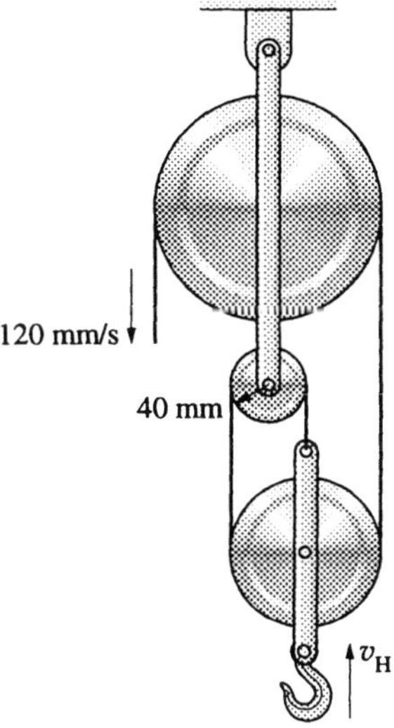

6.72ps At the instant shown, arm *BC* is sliding at 0.75 m/s relative to the sleeve at *C* and arm *AB* is rotating in the counterclockwise direction. Determine the angular velocity of arm *AB*.

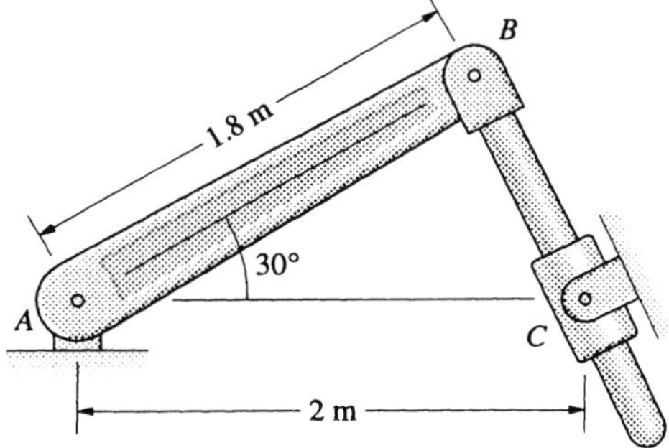

6.73ps At the instant shown, the athlete's shoulder joint A is stationary, $w_{AB} =$ - 1 rad/s, and $w_{BC} = 1$ rad/s The mass $m = 8$ kg. What is the tension in the cable at the present instant?

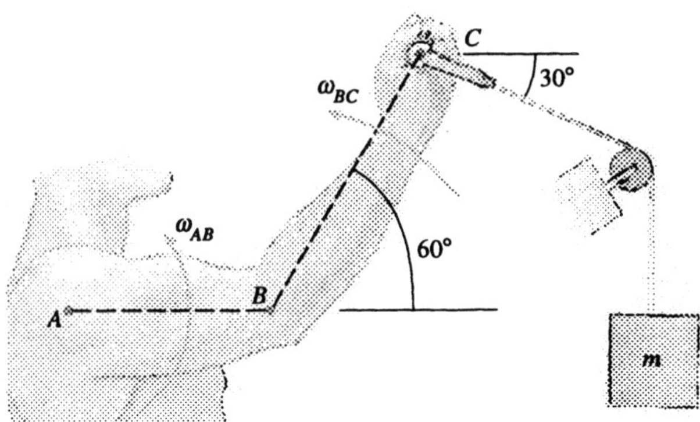

6.74ps The position of the stationary buoy A relative to the ship-fixed coordinate system is $\mathbf{r}_{A/B} = 400\mathbf{i} + 250\mathbf{j}$ (m). The ship is moving in the y axis direction at a steady speed of 10 knots and is turning to the left at a constant rate of 2° per second. The ship uses its radar to measure the velocity and acceleration of the buoy relative to the ship-fixed coordinate system at this instant. What are they?

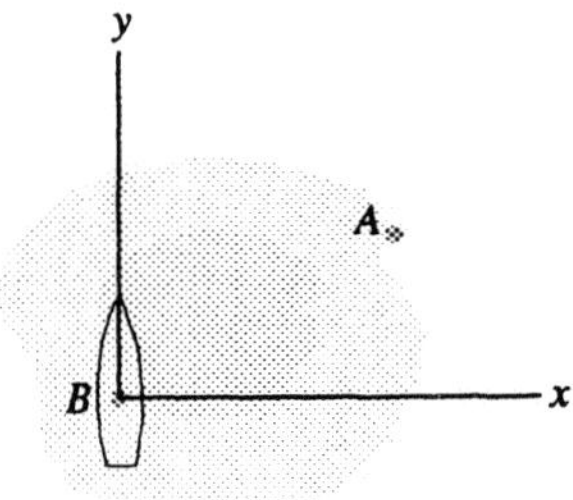

Chapter 7: Two-Dimensional Dynamics of Rigid Bodies

7.1ps The mass of the Boeing 747 is 250 Mg (megagrams). Suppose that the aircraft reverses thrust during a takeoff run to abort the takeoff. The thrust T is opposite to the direction shown and its magnitude is 300 kN. Determine the deceleration of the airplane caused by the reversed thrust and the normal forces exerted on its wheels at A and B. Neglect the horizontal forces exerted on its wheels.

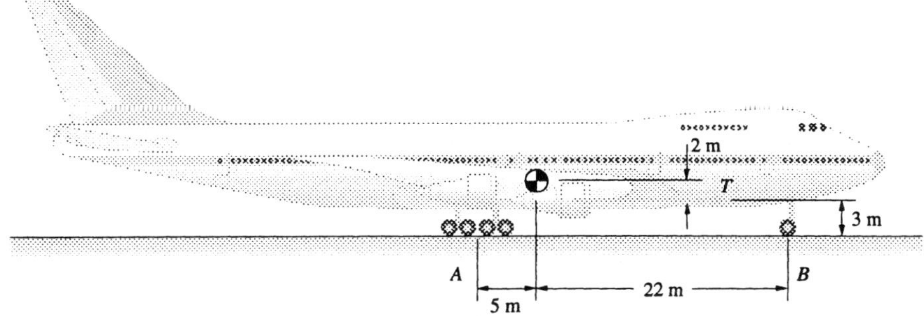

7.2ps In Problem 7.1ps, suppose that the pilot applies the airplane's brakes in addition to reversing the thrust. If the effective coefficient of friction due to the brakes being applied to the rear wheels is 0.6, and the nose wheel does no braking, determine the airplane's deceleration and the normal forces exerted on its wheels at A and B.

7.3ps The motor is being used to pull the 100-lb crate up the smooth ramp when the power fails, reducing the torque M to zero. The mass moment of inertia of the drum on which the cable is wound is 3 slug-ft^2. What is the magnitude of the crate's acceleration down the ramp after the power fails?

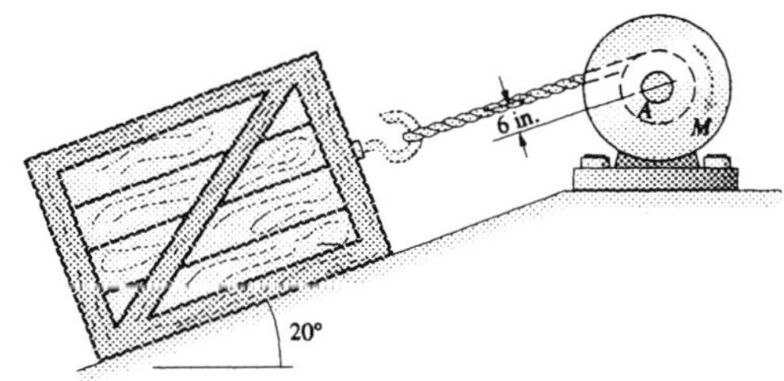

7.4ps In Problem 7.3ps, what torque M would the motor need to exert to cause the crate to accelerate down the smooth ramp at 0.5 m/s^2.

7.5ps The mass of the slender bar is 2 kg and its length is 1.5 m. The bar is inhomogeneous and its center of mass, G, is 0.5 m from the bottom of the bar [$I_G = (2/27)mL^2$]. The angular velocity ω is zero, the angle $\theta = 30°$, and the floor and wall are smooth. What is the bar's angular acceleration?

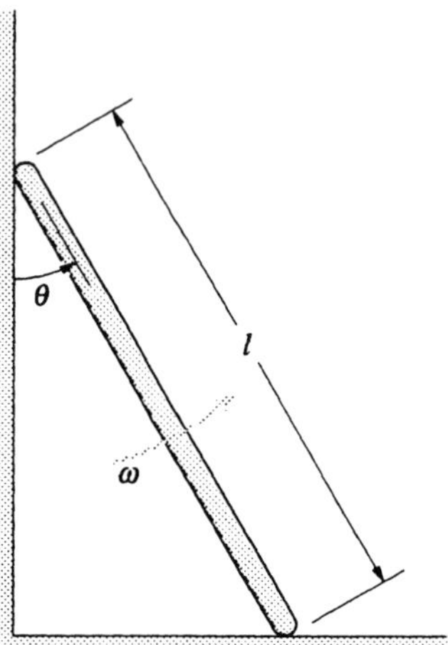

7.6ps In Problem 7.5ps, what is the bar's angular acceleration if the angular velocity $\omega = 0.5$ rad/s counterclockwise and $\theta = 30°$.

7.7ps The length of the homogeneous 2-kg bar is $l = 1.2$ m. The bar starts from rest and falls from the vertical position shown under the influence of gravity. Let θ be the angle between the bar and the vertical. Draw a graph of the bar's angular velocity as a function of θ for $0 \leq \theta \leq 90°$.

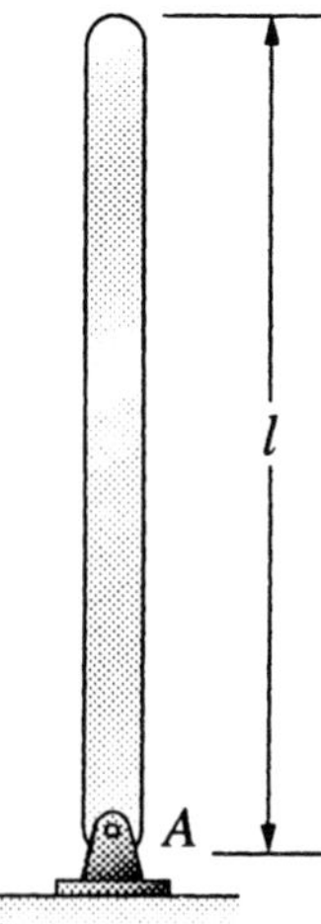

7.8ps The uniform 4-kg disk has a radius $R = 0.5$ m. The angle $\beta = 27°$. The coefficients of static and kinetic friction between the disk and the inclined surface are 0.6 and 0.4, respectively. If the disk is released from rest, what are its angular acceleration and the acceleration of its center?

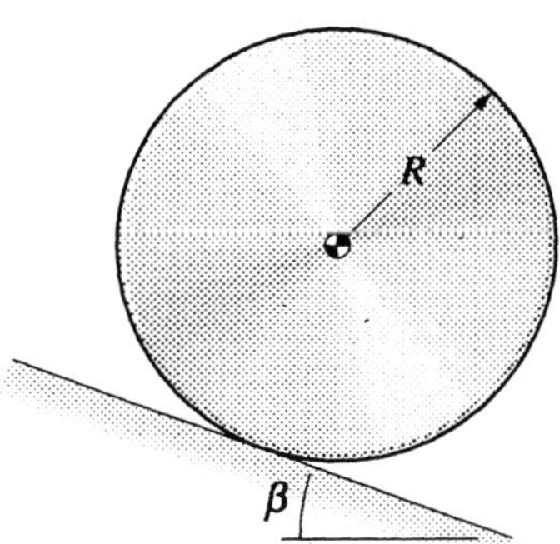

7.9ps Solve Problem 7.8ps with the angle β=45°.

7.10ps The mass of the bicycle and rider is 75 kg. The dimensions $b = 615$ mm, $c = 445$ mm, and $h = 985$ mm. The rider is moving at 5 m/s when she engages only the rear wheel brake, locking the wheel. If μ_k=0.4, what is her stopping distance?

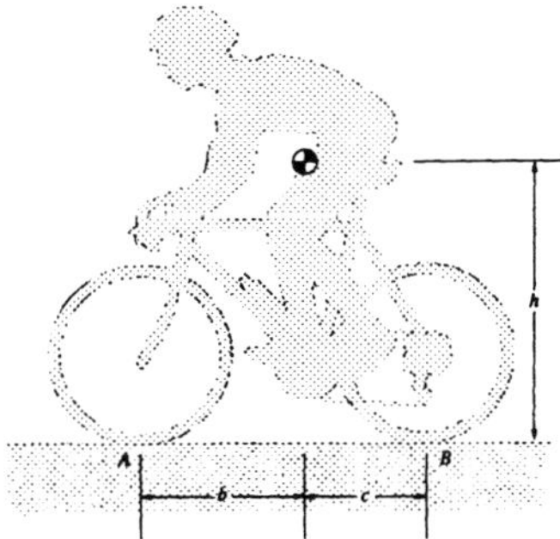

7.11ps Solve Problem 7.10ps assuming that she engages only the front wheel brake, locking the front wheel. Does the rear wheel leave the ground?

7.12ps Solve Problem 7.10ps assuming that she engages both rear and front wheel brakes. Does the rear wheel leave the ground?

7.13ps In Problem 7.10ps, determine the minimum value of μ_k that will cause the rear wheel to leave the ground if only the front wheel is locked during braking.

7.14ps The 14,000-lb airplane shown catches a cable with its tailhook while traveling horizontally at 140 mph. The cable exerts a constant force $F = 180{,}000$ lb in the direction shown to decelerate the airplane. How long does it take the airplane to stop, and what normal forces are exerted on the landing gear at A and B while it is decelerating?

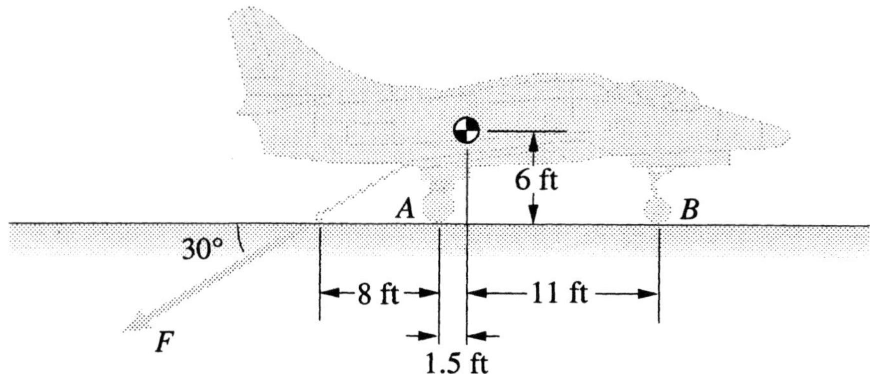

7.15ps In Problem 7.14ps, suppose that the arresting cable breaks 40 ft beyond the point where the tailhook first catches it. What is the speed of the airplane when the cable breaks? Assume that the 30° angle remains constant.

7.16ps The combined mass of the motorcycle and rider is 160 kg and their center of mass is located as shown. The rider is riding at high speed when he sees a hazard on the road ahead and applies the brakes, locking both the front and rear wheels. If $\mu_k = 0.5$, determine the normal forces on the front and rear wheels and the deceleration of the motorcycle

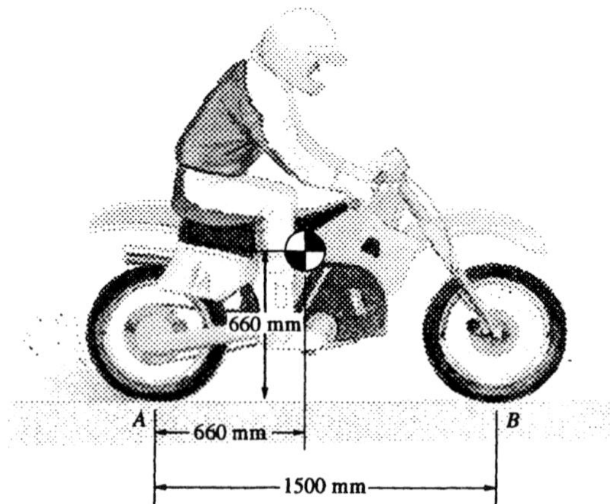

7.17ps The rider in Problem 7.16ps begins to lose control and releases the brake on the front wheel, keeping the rear wheel locked. Determine the normal forces on the front and rear wheels and the deceleration of the motorcycle.

7.18ps The mass moment inertia of the pulley is 0.4 slug-ft^2 and the weight in (b) is 20 lb on the Earth.. Determine the angular acceleration of the pulley in cases (a) and (b) if: (1) the systems are on Earth; (2) the systems are on Mars where the acceleration due to gravity is 13.2 ft/s^2.

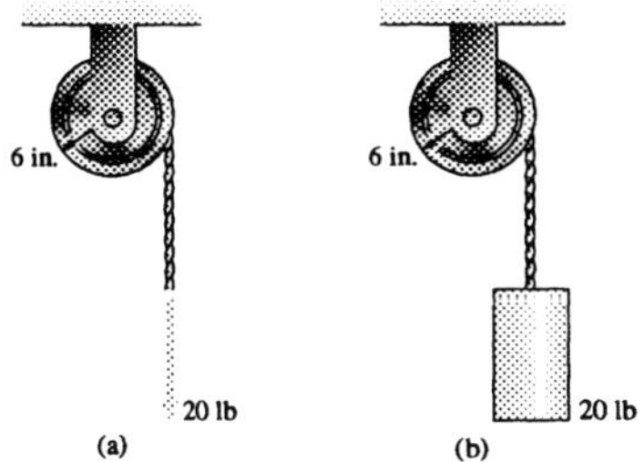

7.19ps The slender bar weighs 10 lb and the disk weighs 20 lb. The coefficient of kinetic friction between the disk and the surface is 0.1. The disk has an initial clockwise angular velocity of 10 rad/s. To cause the disk to stop rotating more quickly, suppose that we consider either increasing the weight of the bar by 10 lb or increasing the weight of the disk by 5 lb. Analyze both possibilities and recommend which option to take.

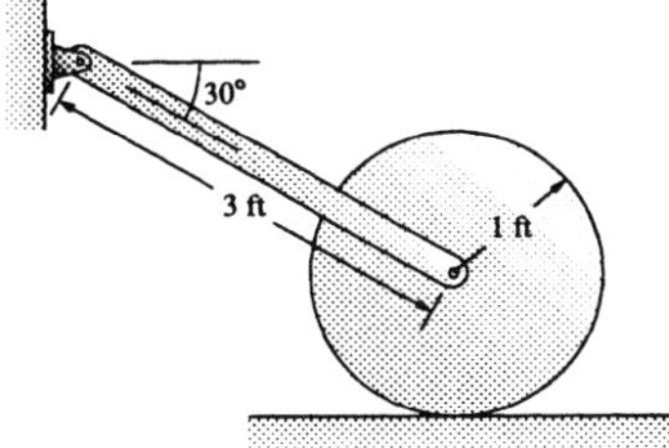

7.20ps Solve Problem 7.19ps assuming that the initial angular velocity of the disk is counterclockwise instead of clockwise.

7.21ps The mass of the bar is $m = 2$ kg and its length is $l = 4$ m. If the bar is released from rest in the horizontal position, draw a graph of its initial angular acceleration as a function of x for $0 \leq x \leq 2$ m.

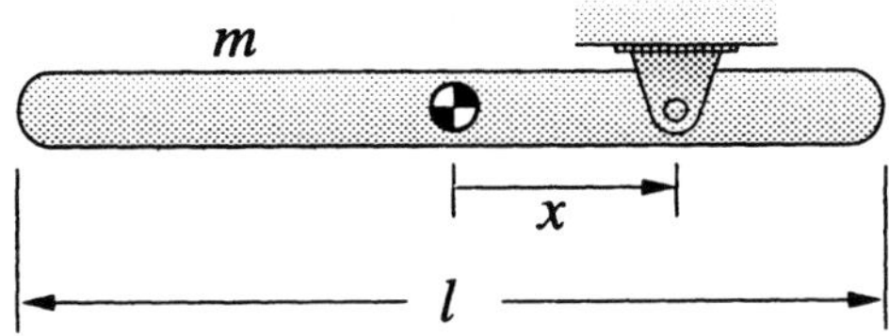

7.22ps Consider the ring and disk shown and also a homogeneous sphere, each of mass $m = 1$ kg and radius $R = 0.2$ m. The mass moment of inertia of the sphere about an axis through its center is $I = (2/5)mR^2$. Each object is placed on a 20° incline and released from rest. Determine the time required for each one to roll 3 meters down the incline.

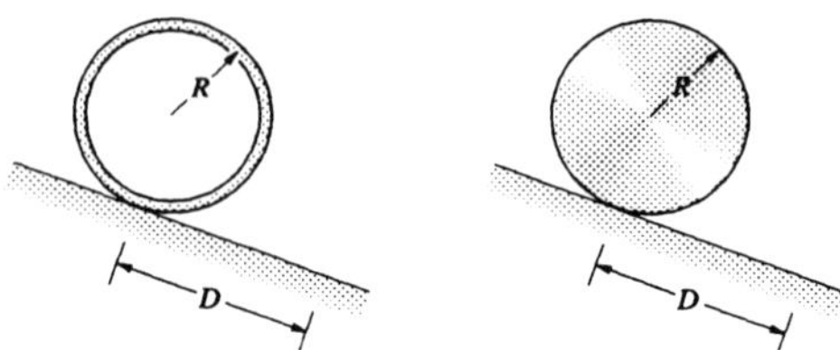

7.23ps The mass moment of inertia of the 40-lb stepped disk is 0.2 slug-ft². If the disk is released from rest in the position shown, what is its angular velocity and what is the tension in the string?

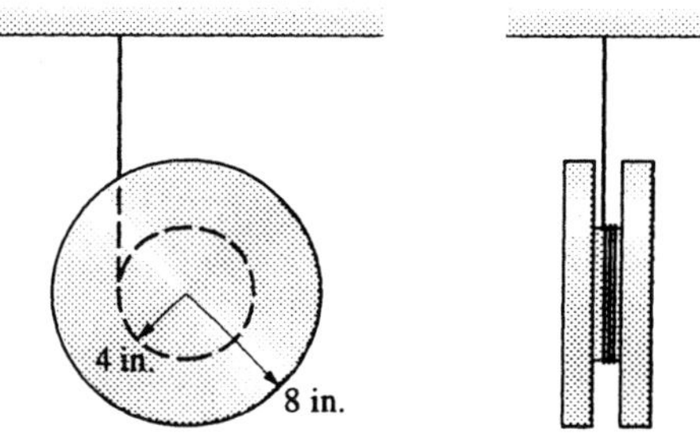

7.24ps The homogeneous sphere $[I = (2/5)mR^2]$ has 200-mm radius and 3-kg mass. At $t = 0$ it is placed on the flat surface with an initial angular velocity $\omega_0 = 10$ rad/s. The coefficient of kinetic friction is 0.4. Draw graphs of the sphere's angular velocity and the velocity of its center as functions of time from $t = 0$ until the time at which the center of the sphere reaches its maximum velocity.

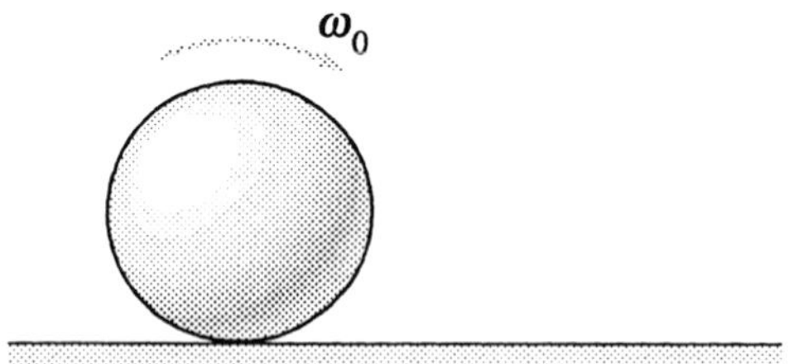

7.25ps The couple M causes the 60-lb slender bar to pull the 80-lb crate to the left on the smooth surface. At the instant shown, the velocity of the crate is 20 ft/s to the left and the couple M suddenly reduces to zero. Determine the acceleration of the crate and the angular acceleration of the bar at that instant.

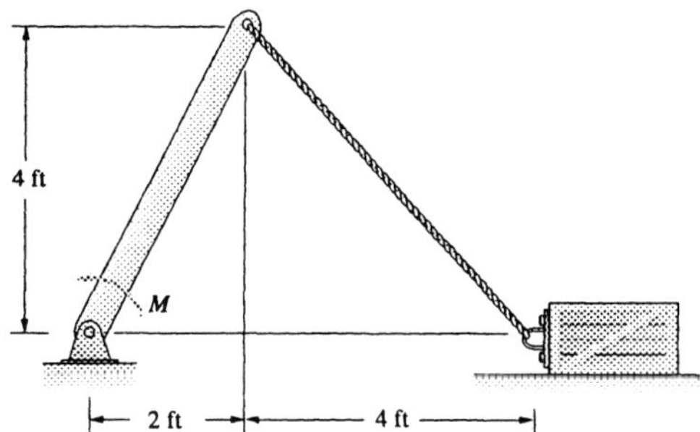

7.26ps Bar *AB* weighs 10 lb and bar *BC* weighs 6 lb. The surface at *C* is smooth. At the instant shown, the bars are stationary and an upward 12-lb force is applied at *B*. At that instant, what are the angular accelerations of the bars?

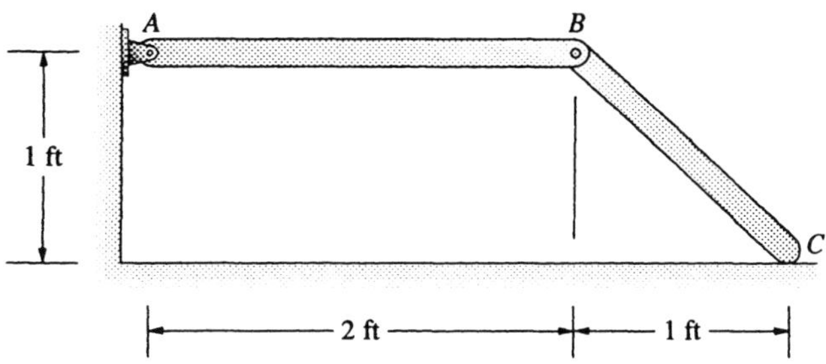

7.27ps The rocket is 80 ft long and weighs 100,000 lb. It is accelerating upward at 1.5 *g*'s. Modeling the rocket as a homogeneous bar, draw a graph of the internal axial force as a function of the axial distance *x* measured from the bottom of the rocket for $0 < x < 80$ ft.

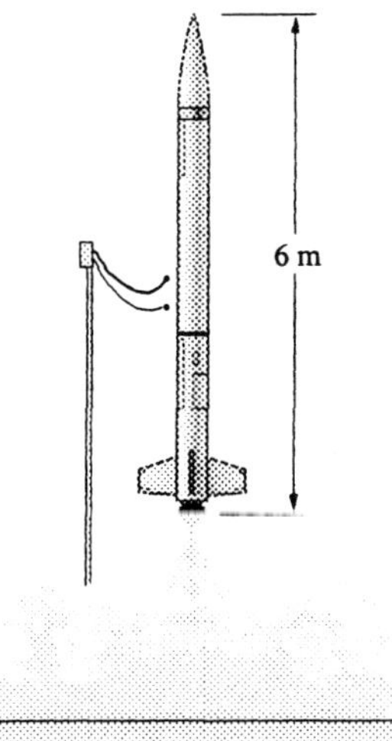

7.28ps A 2-kg disk has a radius of 0.5 m. Its mass moment of inertia about the axis through its center that is perpendicular to the disk is 0.3 N-m2. Is this a homogeneous disk?

7.29ps The thin homogeneous triangular plate is of uniform thickness and mass m. Determine the mass moments of inertia of the plate about axes x', y', z' that are parallel to the x, y, z axes and whose origin is at the center of mass.

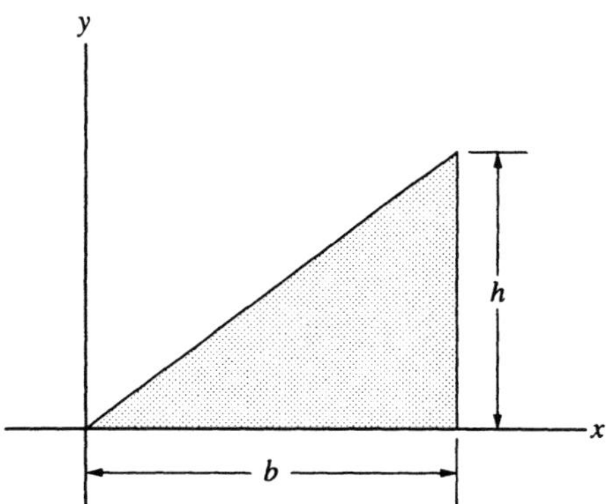

7.30ps The object shown consists of a 3-kg bar welded to a 2-kg thin disk. The axis L is perpendicular to the bar, perpendicular to the disk, and passes through the center of mass of the object. Determine the mass moment of inertia of the object about the axis that is parallel to L and passes through the center of the disk.

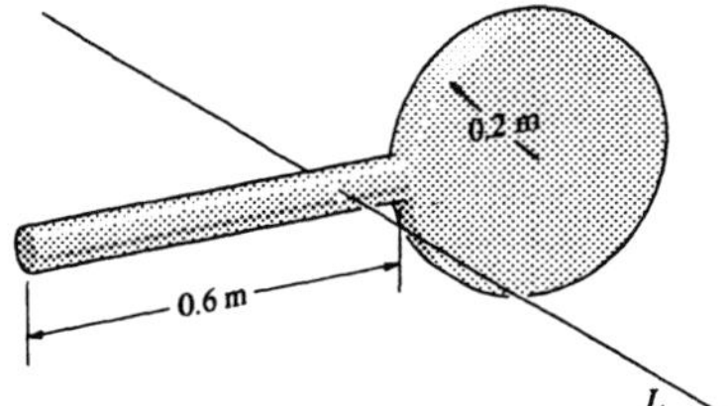

7.31ps The mass moment of inertia of a homogeneous sphere of mass m and radius R about an axis through its center is $I = (2/5)mR^2$. Use this result to determine the mass moment of inertia about an axis through the center of a hollow sphere of the same material with outer radius R and inner radius R_2. Strategy: Express the mass moment of inertia of the solid sphere in terms of the mass density ρ of the material.

7.32ps The thin circular ring of mass m and radius R was formed by bending a straight uniform bar into a circle. Suppose that a straight section of identical bar with length $2R$ is placed across the circle along the axis L and welded to it at both ends. Sketch the new object. Determine its mass moment of inertia: (a) about the axis L; (b) about the axis through the center of mass of the object that is perpendicular to the ring; (c) about the axis through the plane of the ring that is perpendicular to *L*.

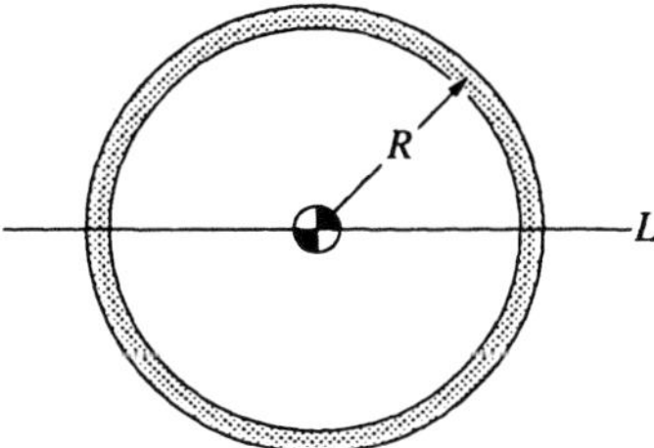

7.33ps Two homogeneous slender bars, each of mass m and length l, are welded together to obtain the object shown. Use the parallel-axis theorem to determine the mass moment of inertia of the object about the axis perpendicular to the two the bars that passes through the point at which the bars are welded.

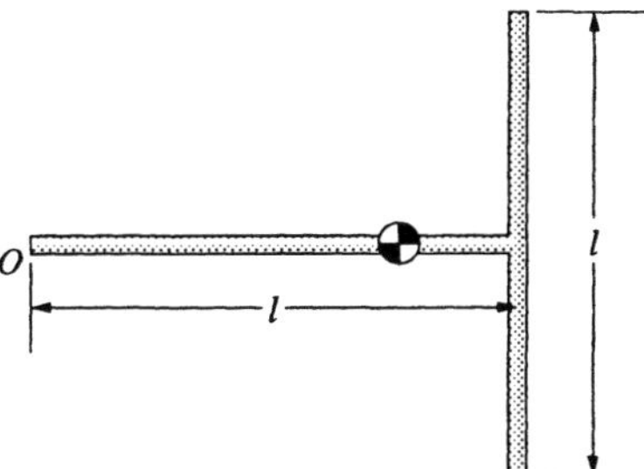

7.34ps Suppose that a 10-kg straight bar that is 2.5 m in length is welded to the 20-kg bar shown so that it connects the free ends of the bar. (The new bar lies along the x axis). Determine the mass moment of inertia of the new composite bar about the axis through its center of mass that is parallel to the z axis.

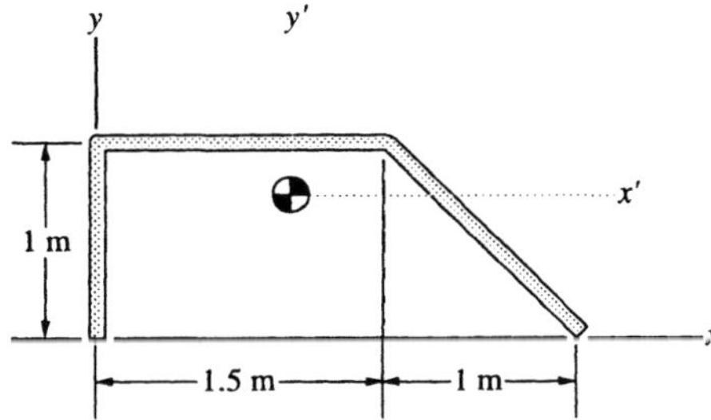

7.35ps The homogeneous disk has a mass of 3 kg and a radius of 0.5 m. If $\mu_s = 0.5$, what is the largest angle β for which the disk, released from rest, will roll down the incline without slipping? Refer to the figure in problem 7.8ps.

7.36ps The pilot of the 1000-lb experimental airplane is rolling down the runway at 65 mi/hr when he discovers a problem and turns the engine off. The thrust T is reduced to zero. If the brakes affect the rear wheels only (the nose wheel does not brake) and the effective coefficient of friction is 0.4, determine the distance required for the airplane to stop and the normal force exerted on the nose wheel while it decelerates.

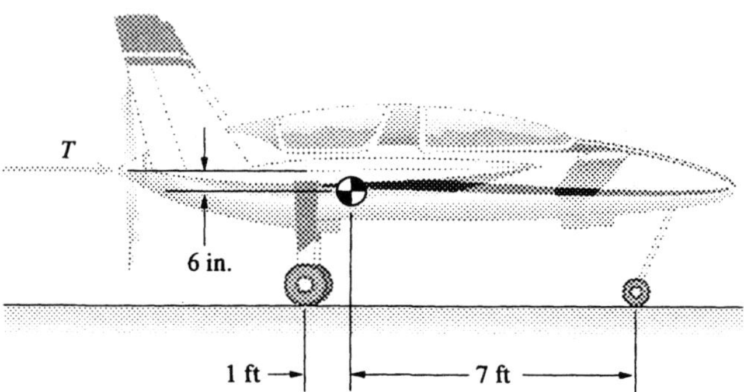

7.37ps After the incident described in Problem 7.36ps, the pilot decides to calculate a "safe" runway length for his airplane. He defines this to be the sum of three distances: (1) the distance required to accelerate from rest to a speed of 80 mi/hr with a thrust of 300 lb (neglect other horizontal forces); (2) a decision distance during which he decides whether to take off or attempt to stop. He assumes this to be the distance traveled in 0.5 seconds at a constant speed of 80 mph; and (3) a braking distance (rear wheels only) with no thrust and an effective coefficient of friction of 0.3. What is the safe runway length? The center of mass of the airplane is 3 ft abvoe the ground.

7.38ps The three pulleys can turn freely on their pin supports. Their mass moments of inertia are $I_A = 0.002$ kg-m^2, $I_B = 0.036$ kg-m^2, and $I_C = 0.032$ kg-m2. The system starts from rest at $t = 0$. Determine the angular velocity of pulley A at $t = 2$s if a constant counterclockwise torque of 5 N-m is applied: (a) to pulley A; (b) to pulley B; (c) to pulley C.

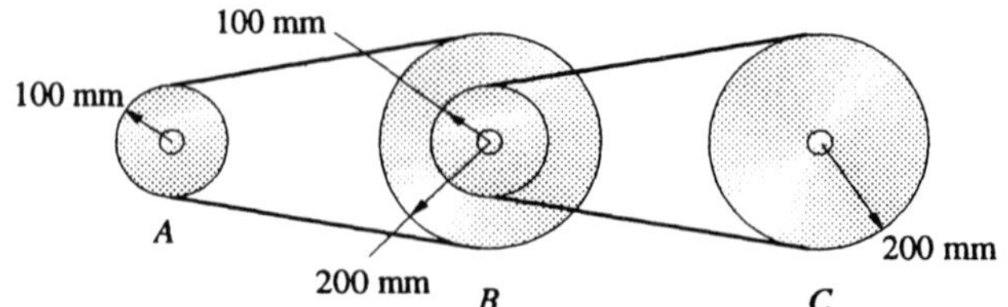

7.39ps The 2-kg slender bar AB is 3 ft long. It is pinned to the cart at A and leans against it at B. Find the acceleration a of the cart for which the normal force exerted on the bar at B is twice the weight of the bar.

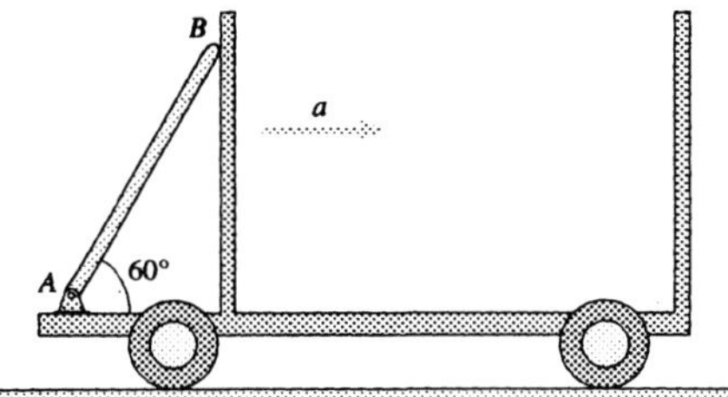

7.40ps As a new engineer at a tire factory, you are asked to develop a method to determine the mass moments of inertia I of tires about the axis through their centers perpendicular to the plane of the tire. The tires have a 195-mm radius and a mass of 6.35 kg. The method you suggest involves rolling the tires from rest down the 15° ramp and measuring the time required to roll 5 m. Draw a graph of I as a function of the time required for a tire to roll 5 m down the ramp. Consider values of I in the range $(1/2)mR^2 \leq I \leq mR^2$.

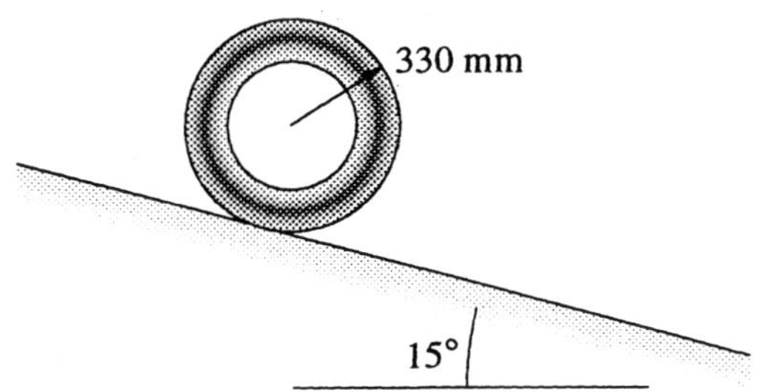

7.41ps The mass moment of inertia of the disk is 0.2 kg-m^2. Suppose that the system is set up on the Earth and also on the Moon where the acceleration due to gravity is 1.62 m/s^2. For each location, determine the smallest coefficient of friction between the rope and the disk for which the rope will not slip when the system is released from rest. If the values are not the same, explain why.

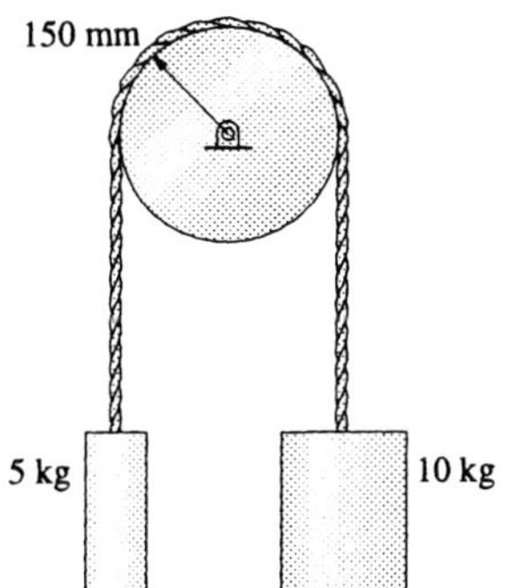

7.42ps The masses of the slender bars *BC* and *CDE* are 3 kg and 6 kg, respectively, and a 5-kg point mass is attached to bar *CDE* at *E*. Bar *AB* rotates with a constant counterclockwise angular velocity of 10 rad/s. Determine the components of force exerted on bar *BC* by the pins at *B* and *C* at the instant shown.

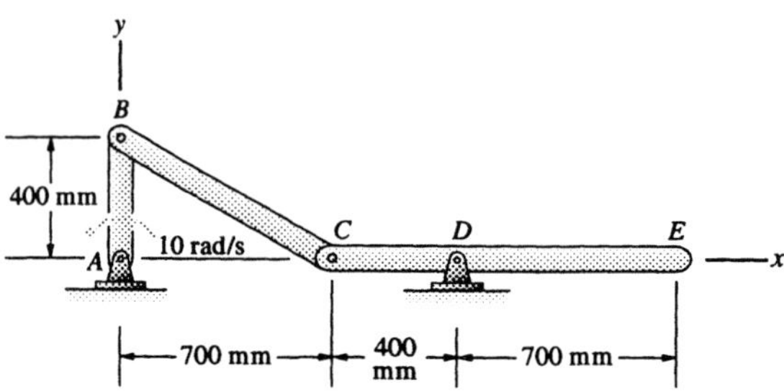

7.43ps In Problem 7.42ps, the slender bar *AB* has a mass of 2 kg. What torque must be applied at *A* to cause it to rotate with constant angular velocity at the instant shown?

Chapter 8: Energy and Momentum in Planar Rigid-Body Dynamics

8.1ps The disk, of mass m and moment of inertia I, is released from rest on the incline. The coefficient of static friction between the disk and the surface is μ_s. Determine the maximum angle β for which the disk will roll without slipping. Assuming that β has this value and that the disk is released from rest, use work and energy to determine the disk's angular velocity after it has rolled a distance b.

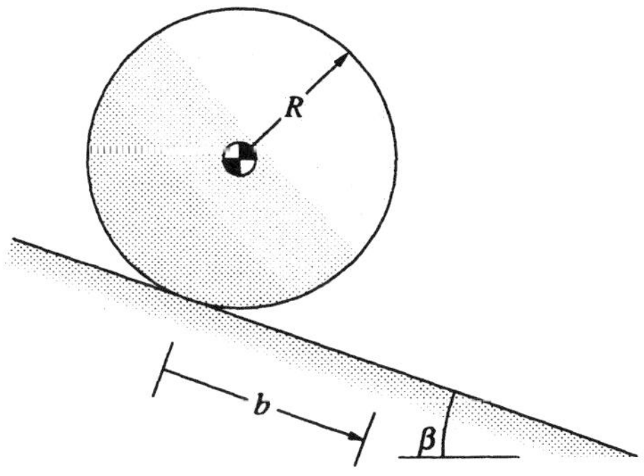

8.2ps A main landing gear wheel of the Boeing 747 has a mass moment of inertia of 17 slug-ft^2 and a 2-ft radius. On landing, the aircraft is moving at 240 ft/s and the wheel touches the ground with no angular velocity. How much work is done on the wheel by friction from the time it touches the ground until it is rolling?

8.3ps The L-shaped object consists of two 1-m, 5-kg bars welded together. It is released from rest in the position shown. Draw a graph of the angular velocity as a function of the angle θ between the bar pinned at A and the horizontal over the range $0 \le \theta \le 90°$. Explain why the maximum angular velocity occur before θ is 90°.

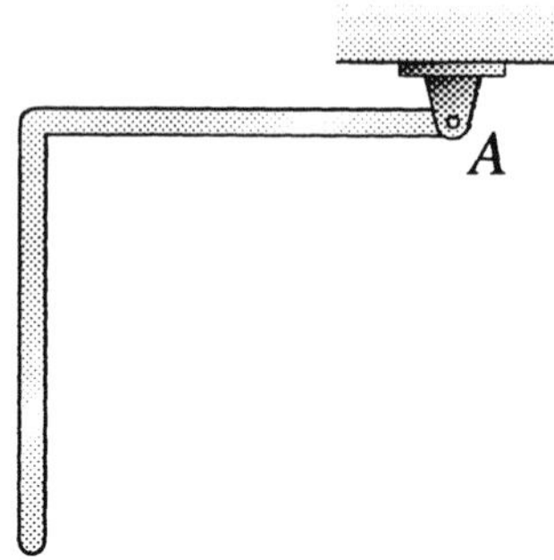

8.4ps The homogeneous sphere is moving horizontally with velocity $v_A = 10$ m/s when it strikes the stationary uniform slender bar. The mass of the sphere is 1 kg, the length of the bar is 2 m, and the mass of the bar is 4 kg. The coefficient of restitution between the bar and sphere is 0.8. Draw a graph of the bar's angular velocity after impact as a function of the distance h for $0.1 \le h \le 1.9$ m.

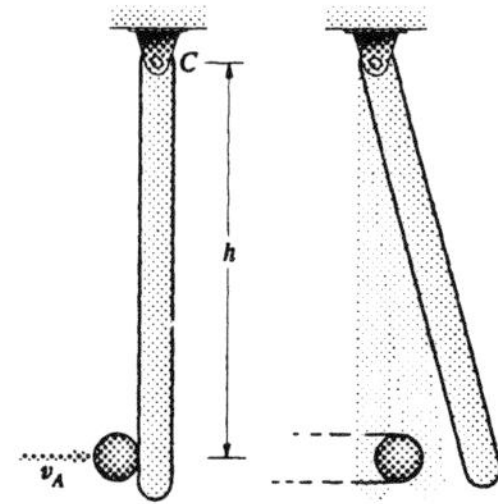

8.5ps Assume that the impact in Problem 8.4ps lasts 0.05 s. Draw a graph of the average horizontal force exerted on the bar by the support C on the bar as a function of the distance h for $0.1 \le h \le 1.9$ m.

8.6ps The 40-lb homogeneous cylindrical disk rolls to the step at 10 ft/s and remains in contact with the step (which is 6 in. in height in the figure) while rolling up onto it. What is the tallest step that the wheel can roll up onto at this speed?

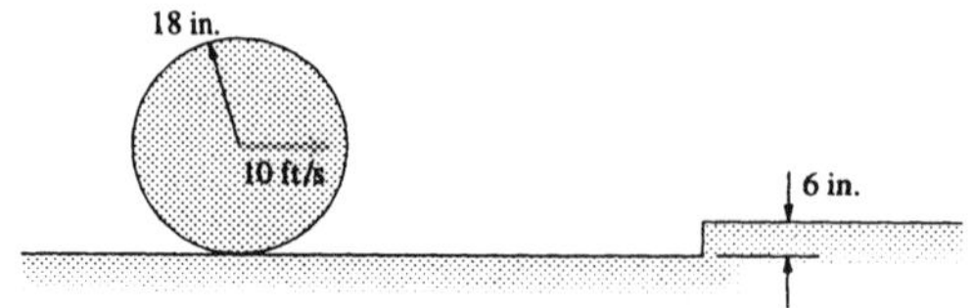

8.7ps In problem 8.6ps, what is the tallest step the wheel can roll up onto if the wheel can be moving with any speed? Does your answer seem to contradict your intuition? What is unrealistic about the analysis?

8.8ps Each slender bar is of mass m and length l. The non-rotating bar A is moving with velocity v_0 when it strikes the stationary bar B. The coefficient of restitution $e = 1$. Determine the velocities of the centers of mass of bars A and B and their angular velocities after they collide.

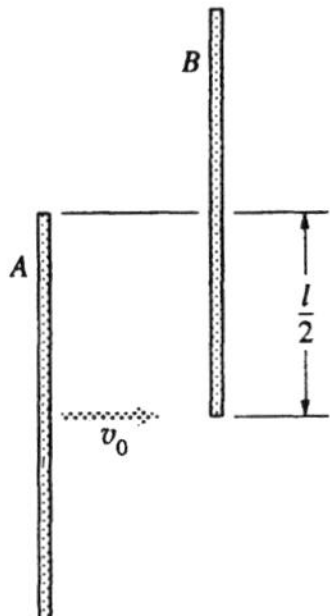

8.9ps In Problem 8.8ps, assume that the stationary bar B is a distance l/6 above the position shown, so that the length of the overlap of the two bars when A strikes B is l/3. If the bars adhere after impact, determine the velocities of the centers of mass of the bars and their angular velocities after they collide. The length l of the bar is 2 m, $v_0 = 1$ m/s, and each bar has a mass of 2 kg.

8.10ps Work problem 8.8ps with $l = 6$ m, $v_0 = 3$ m/s, and $m = 5$ kg.

8.11ps The 10-lb slender bar falls from rest in the vertical position and strikes the smooth projection at B. The coefficient of restitution is 0.7 and the duration of the impact is 0.05s. Determine the average force exerted on the bar at B during the impact if $b = 2$ ft.

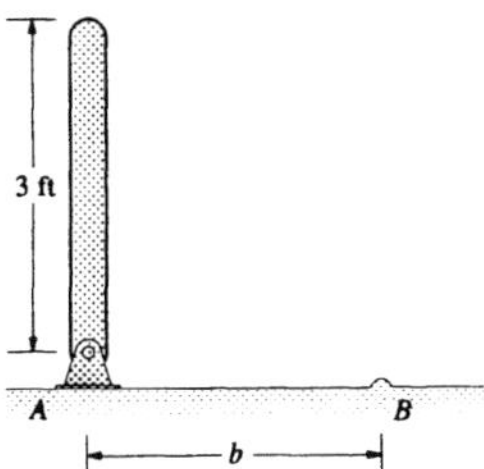

8.12ps In Problem 8.11ps, draw a graph of the average force exerted on the bar at B during the impact as a function of the distance b for $0.5 \leq b \leq 2.5$ ft.

8.13ps The homogeneous disk has a radius $R = 0.5$ m and a mass of 3 kg. The angle $\beta = 15°$ and the angular velocity $\omega = 1.5$ rad/s. Determine the disk's angular velocity ω' immediately after it comes into contact with the incline. Assume that there is no slipping and that the disk remains in contact with the surface.

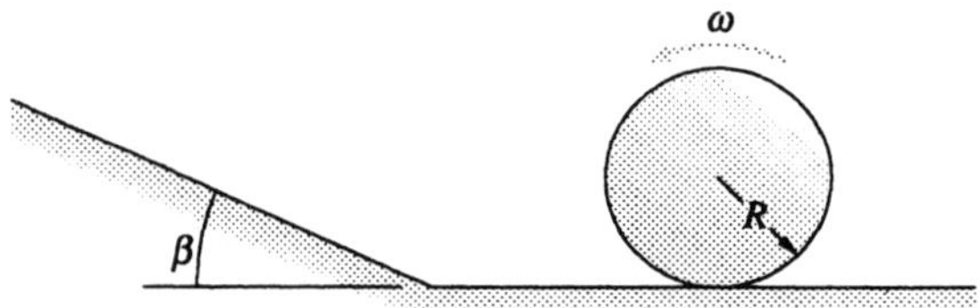

8.14ps In Problem 8.13ps, draw a graph of the disk's angular velocity ω' immediately after it comes into contact with the incline as a function of the angle β for $0 \le \beta \le 35°$.

8.15ps The 1-kg sphere A is moving at 2 m/s when it strikes the stationary 2-kg slender bar B. Let d be the distance from the end of the bar at which A strikes the bar (which is 400 mm in the figure). If the coefficient of restitution is 0.7, draw a graph of the velocity of the top of the bar just after impact as a function of d for $0 \le d \le 1.5$ m. (Comment on your results with respect to the behavior of a baseball bat when a ball is struck.)

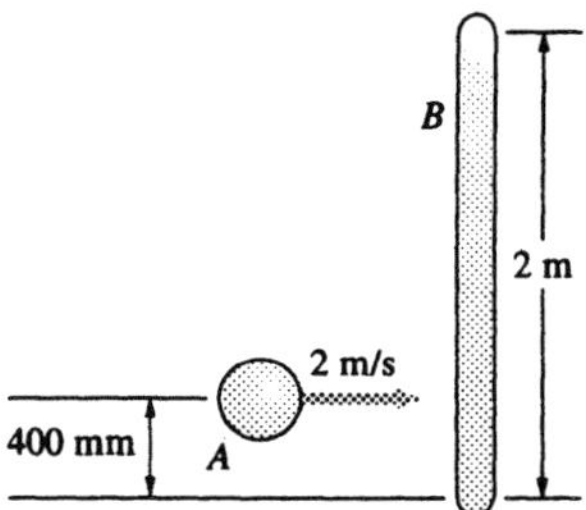

8.16ps The 2-kg slender bar with a length of 1 m is released from rest in the position shown with $h = 1$ m. The coefficient of restitution is 0.4. Draw a graph of the bar's angular velocity just after it strikes the floor as a function of the angle θ for $0 \le \theta \le 80°$.

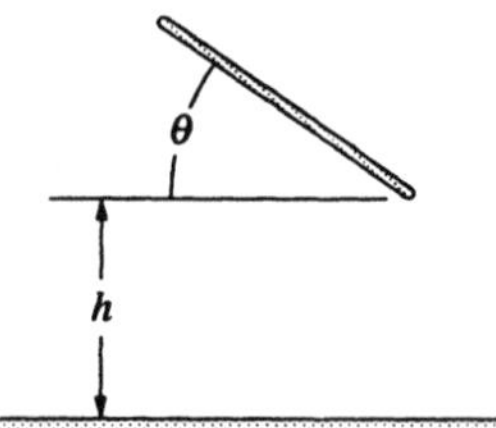

8.17ps The slender bar in Problem 8.16ps is released in the position shown with $\theta = 30°$. Draw a graph of the bar's angular velocity just after it strikes the floor as a function of the coefficient of restitution for $0 \le e \le 1$.

Chapter 9: Three-Dimensional Kinematics and Dynamics of Rigid Bodies

9.1ps Let O be a fixed reference point and let P and Q be points of a rigid body. At the present instant, the angular velocity of the rigid body is $\omega = 3\mathbf{i} - 2\mathbf{j} + 6\mathbf{k}$ (rad/s), the position and velocity of point P relative to O are $\mathbf{r}_P = 16\mathbf{i} - 2\mathbf{j} + 7\mathbf{k}$ (m), $\mathbf{v}_P = 2\mathbf{i} + 2\mathbf{j} - 3\mathbf{k}$ (m/s), and the position of point Q relative to point P is $\mathbf{r}_{Q/P} = -3\mathbf{i} + 6\mathbf{j} - 2\mathbf{k}$ (m). Determine the velocity $\mathbf{v}_Q$ of point Q relative to O and the velocity $\mathbf{v}_{Q/P}$ of point Q relative to point P.

9.2ps At the present instant the angular acceleration of the rigid body in Problem 9.1ps is $\alpha = 2\mathbf{i} - 3\mathbf{j} + 4\mathbf{k}$ (rad/s^2) and the acceleration of point P relative to O is $\mathbf{a}_P = \mathbf{i} + 3\mathbf{j} - 6\mathbf{k}$ (m/s^2). What is the acceleration $\mathbf{a}_Q$ of point Q relative to O?

9.3ps Let O be a fixed reference point and let A and B be points of a rigid body. At the present instant, the angular velocity of the rigid body is $\omega = 14\mathbf{i} - 3\mathbf{j} + 7\mathbf{k}$ (rad/s), the position and velocity of point A relative to O are $\mathbf{r}_A = 16\mathbf{i} + 5\mathbf{j} + \mathbf{k}$ (ft), $\mathbf{v}_A = \mathbf{i} + 2\mathbf{j} + 5\mathbf{k}$ (ft/s), and the position of point B relative to point A is $\mathbf{r}_{B/A} = -3\mathbf{i} + 6\mathbf{j} - 2\mathbf{k}$ (ft). Determine the velocity v_B of point B relative to O and the velocity $\mathbf{v}_{B/A}$ of point B relative to point A.

9.4ps At the present instant the angular acceleration of the rigid body in Problem 9.3ps is $\alpha = -4\mathbf{i} += \mathbf{j} + 5\mathbf{k}$ (rad/s^2) and the acceleration of point A relative to O is $\mathbf{a}_A = 2\mathbf{i} + 9\mathbf{j} - \mathbf{k}$ (ft/s^2). What is the acceleration $\mathbf{a}_B$ of point B relative to O?

9.5ps A light on the airplane's right wingtip is located at (1, 24, −2) ft relative to the airplane's center of mass. A light on the left wingtip is at (1, −24, −2) ft. The rate gyros on the airplane measure its angular velocity to be $\omega = 2\mathbf{i} + 0.4\mathbf{j} + 0.1\mathbf{k}$ (rad/s). Determine: (a) the velocities of the left and right wingtip lights relative to the center of mass; (b) the velocity of the left wingtip light relative to the right wingtip light.

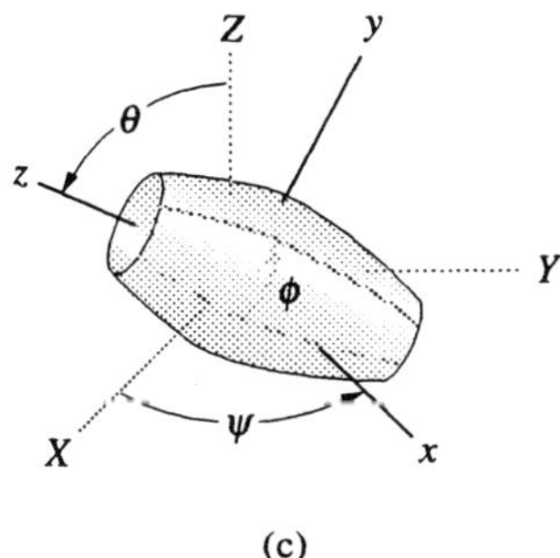

(c)

9.6ps In Problem 9.5ps, the airplane's angular acceleration, measured by an onboard inertial platform, is $\alpha = -0.5\mathbf{i} + 0.1\mathbf{j} + 0.1\mathbf{k}$ (rad/s^2). Determine:

(a) the acceleration of each wingtip light relative to the center of mass;
(b) the acceleration of the left wingtip light relative to the right wingtip light.

9.7ps The airplane shown in Problem 9.5ps undergoes the following series of maneuvers. What is the airplane's angular velocity vector during each maneuver?
(a)The airplane executes a steady roll (rotation about the x axis with constant angular velocity). When the pilot initiates the roll the wingtip to his right moves downward, and the airplane completes one revolution in one second.
(b)The airplane executes a steady loop (rotation about the y axis with constant angular velocity). When the pilot initiates the loop the nose moves upward, and the airplane completes one revolution in 60 seconds.
(c)The airplane executes a steady yaw (rotation about the z axis with constant angular velocity). When the pilot initiates the yaw the nose moves to his left, and the airplane completes one revolution in six minutes.

9.8ps The parallelepiped rotates about the fixed axis *AB*. The magnitude of the angular velocity is 30 rad/s and the magnitude of the angular acceleration vector is 15 rad/s². The angular velocity vector and angular acceleration vector point in the same direction. What is the acceleration of point *C*?

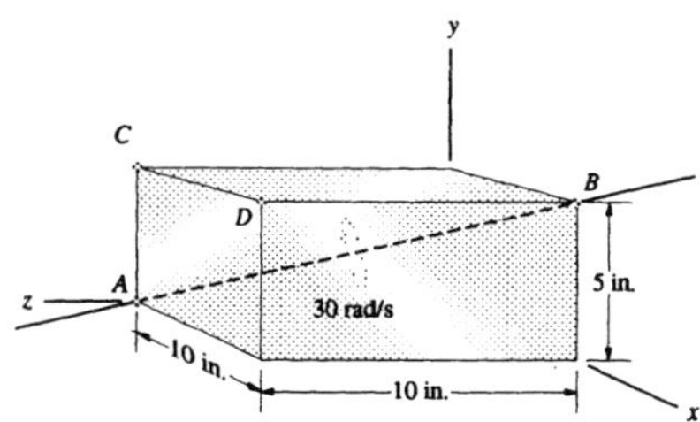

9.9ps The base of the antenna rotates at 1 rad/s. At the present instant the angle $\theta = 30°$, $d\theta/dt = -20°/s$, and $d^2\theta/dt^2 = 10°s^2$. Determine the velocity of the antenna's feed horn, which is located at (5,0,0) m.

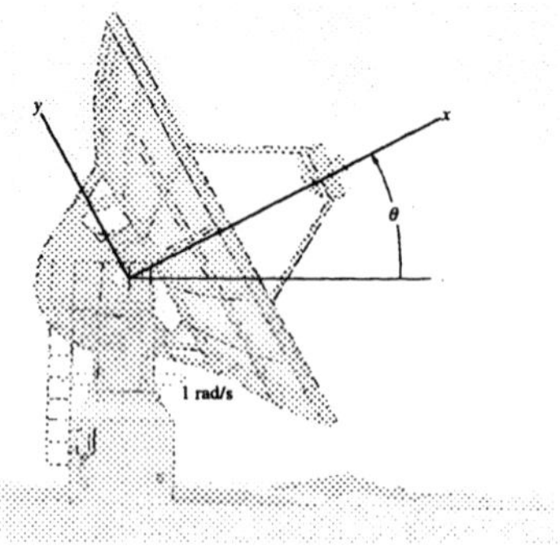

9.10ps In Problem 9.9ps, determine the acceleration of the antenna's feed horn.

9.11ps The circular disk remains perpendicular to the horizontal shaft and rotates relative to it with angular velocity $\omega_d = 4$ rad/s. If $b = 1$ m, $R = 0.5$ m, and $\omega_0 = 6$ rad/s, find the point on the circular disk which has: (a) the maximum velocity magnitude relative to the origin; (b) the minimum velocity magnitude relative to the origin.

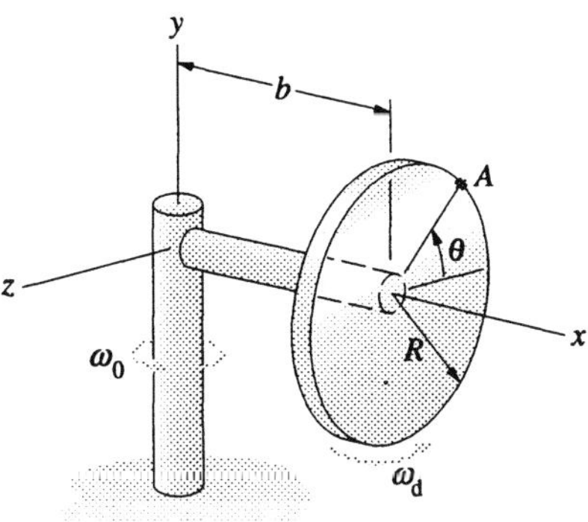

9.12ps Solve Problem 9.11ps with $\omega_d = 6$ rad/s, $b = 1$ m, $R = 0.5$ m, and $\omega_0 = 20$ rad/s.

9.13ps The inertia matrix for a rigid body in terms of a body-fixed coordinate system with its origin at the center of mass is \bdi

[I]	=	7	–1	1	
		–1	1	0	
		1	0	3	kg-m2.

If the rigid body's angular velocity is $\omega = \mathbf{i} + 6\mathbf{j} + \mathbf{k}$ (rad/s), what is its angular momentum about its center of mass?

9.14ps For the inertia matrix in Problem 9.13ps, determine the principal moments of inertia.

9.15ps For the inertia matrix in Problem 9.13ps, determine the components of unit vectors indicating the directions of the principal axes and state which principal moment of inertia is associated with each axis.

9.16ps A rigid body rotates about a fixed point *O*. Its inertia matrix in terms of a body-fixed coordinate system with its origin at *O* is given by

65	–47	17	
–47	9	32	slug-ft2
17	32	165	

If the rigid body's angular velocity is $\omega = 9\mathbf{i} + 2\mathbf{j} + 13\mathbf{k}$ rad/s, what is its angular momentum about *O*?

9.17ps For the inertia matrix in Problem 9.16ps, determine the principal moments of inertia.

9.18ps For the inertia matrix in Problem 9.16ps, determine the components of unit vectors indicating the directions of the principal axes and state which principal moment of inertia is associated with each axis.

9.19ps The mass of the slender homogeneous bar is 6 kg. Determine the bar's inertia matrix in terms of a coordinate system parallel to the coordinate system shown but with its origin at the right hand end of the horizontal bar.

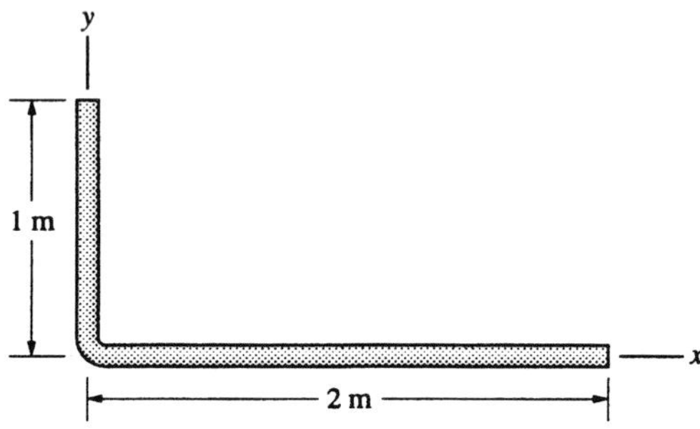

9.20ps For the inertia matrix determined in Problem 9.19ps, determine the principal moments of inertia and the components of unit vectors indicating the directions of the principal axes.

9.21ps The rectangular plate rotates with angular velocity $\omega = 4\mathbf{i} + 6\mathbf{j} - 3\mathbf{k}$ (rad/s). What is its angular momentum about its center of mass?

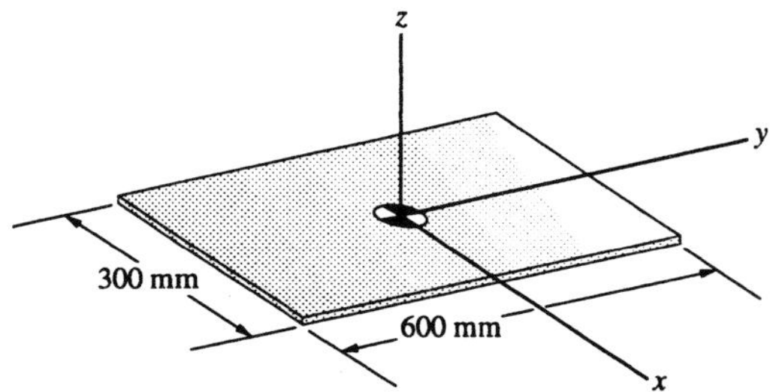

9.22ps The 30-lb thin triangular plate lies in the x-y plane. Determine the plate's inertia matrix in terms of a coordinate system parallel to the coordinate system shown but with its origin at the plate's lower right hand corner.

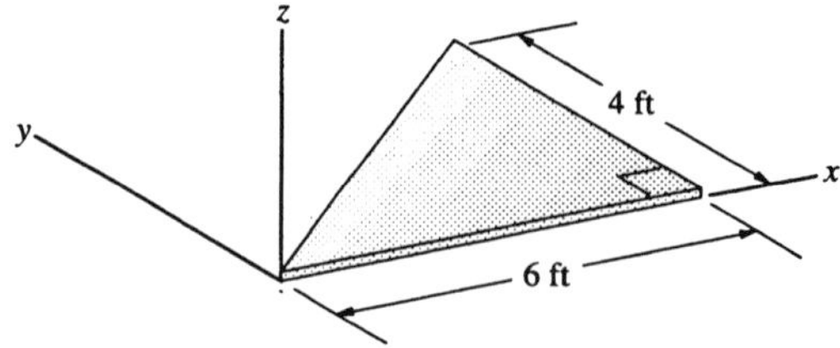

9.23ps For the inertia matrix determined in Problem 9.22ps, determine the principal moments of inertia and the components of unit vectors indicating the directions of the principal axes.

9.24ps Determine the inertia matrix of the 0.6-slug thin plate in terms of a coordinate system parallel to the coordinate system shown but with its origin at the plate's center of mass. Based on your result, what can you say about the plate's principal axes and principal moments of inertia?

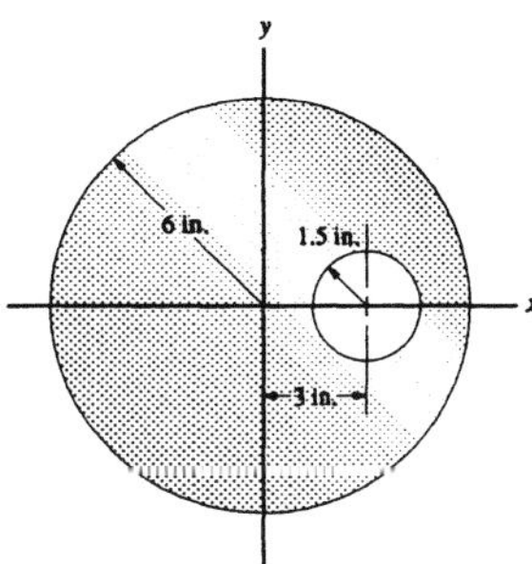

9.25ps The dimension $b = 1$ m. Determine the inertia matrix and the principal moments of inertia of the 4-kg bar.

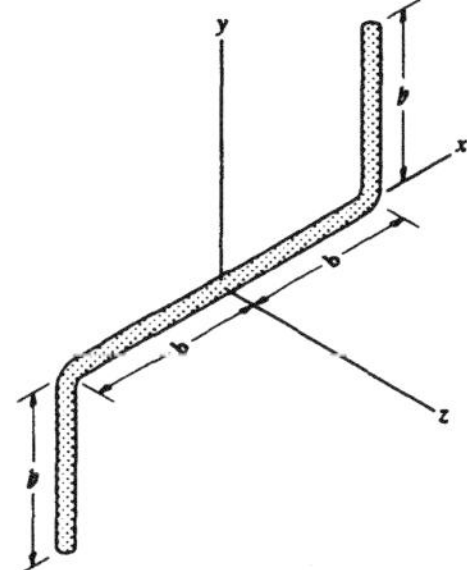

9.26ps For the bar in Problem 9.25ps, determine the components of unit vectors indicating the directions of the principal axes.

9.27ps Determine the inertia matrix for the 8-kg bar in terms of the coordinate system shown.

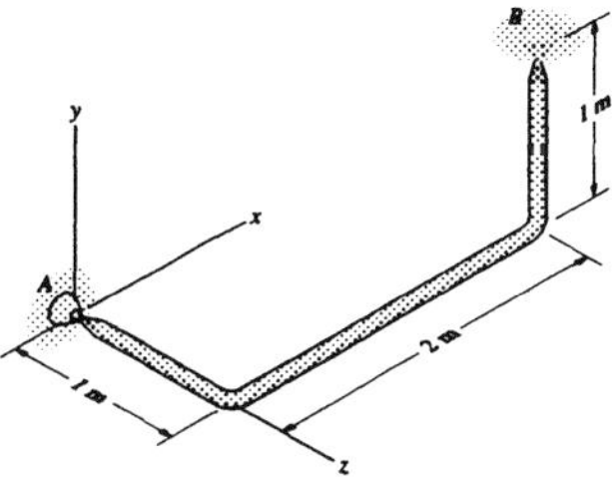

9.28ps Determine the inertia matrix of the bar in Problem 9.27ps in terms of a coordinate system parallel to the coordinate system shown but with its origin at the bar's center of mass.

9.29ps For the inertia matrix determined in Problem 9.28ps, determine the principal moments of inertia and the components of unit vectors indicating the directions of the principal axes.

9.30ps The inertia matrix of a rigid body in terms of a body-fixed coordinate system with its origin at the center of mass G is given by

$$\begin{matrix} 1 & -1 & 1 \\ -1 & 2 & -2 \\ 1 & -2 & 3 \end{matrix} \quad \text{kg-m}^2.$$

Determine the principal moments of inertia.

9.31ps For the inertia matrix in Problem 9.30ps, determine the components of unit vectors indicating the directions of the principal axes and state which principal moment of inertia is associated with each axis.

9.32ps The top is in steady precession with nutation angle $\theta = 10°$ and spin rate $\phi =$ 60 revolutions per second. The mass of the top is 0.01 slug and its moments of inertia are $I_{xx} = 2 \times 10^{-4}$ slug-ft^2 and $I_{zz} = 4 \times 10^{-4}$ slug-ft^2. Determine the top's precession rate φ. The top's center of mass is 2 inches from the support point.

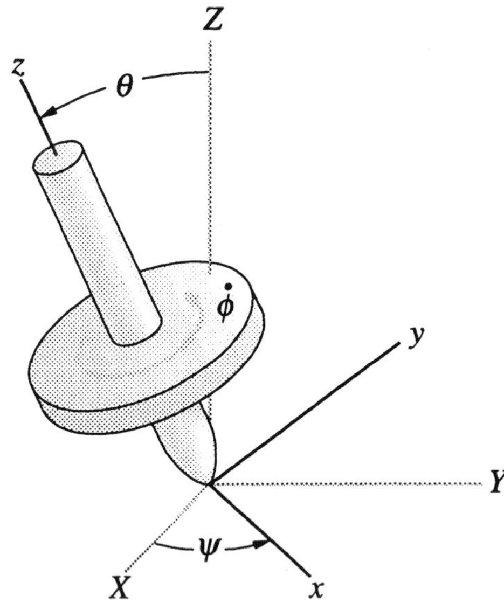

9.33ps The 4-kg rectangular plate is held at O by a robotic manipulator. At the instant shown, the plate's angular velocity and angular acceleration are $\omega = 3\mathbf{i} + 4\mathbf{j}$ (rad/s) and $\alpha = -2\mathbf{i} - 3\mathbf{k}$ (rad/s^2), and the velocity and acceleration point O are $\mathbf{v}_O = \mathbf{i} + 2\mathbf{j} + \mathbf{k}$ mm/s and $\mathbf{a}_O = -0.1\mathbf{i} + \mathbf{j} - 0.5\mathbf{k}$ mm/s^2. Determine the couple exerted on the plate by the manipulator at O.

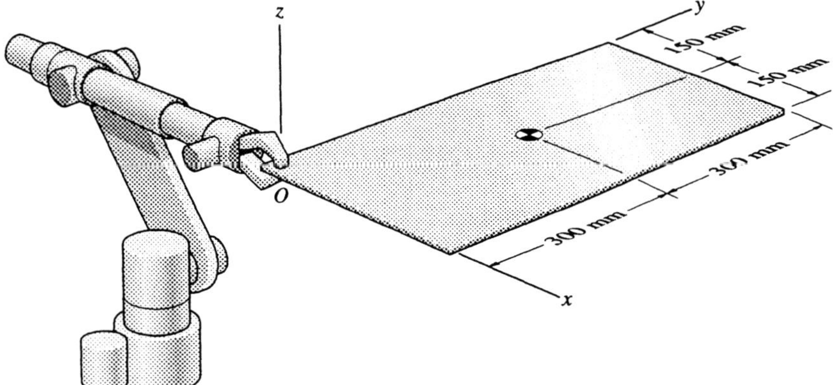

Chapter 10: Vibrations

10.1ps The pulley has radius $R = 1$ ft and a mass of 0.5 slugs. The mass $m = 1$ slug and the spring constant is $k = 60$ lb/ft. If the mass is released from rest with the spring unstretched, determine the position of the mass relative to its initial position as a function of time.

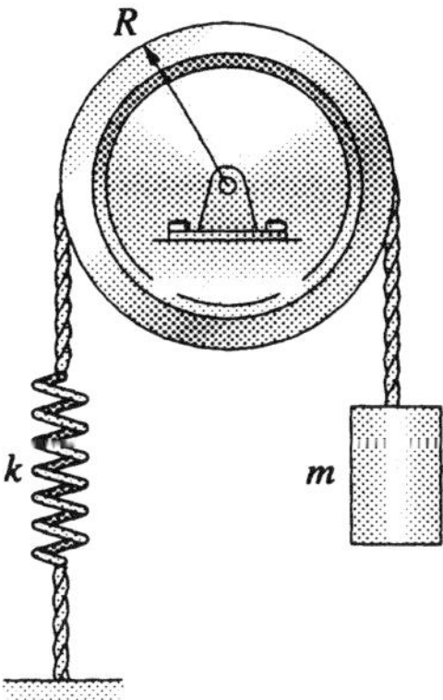

10.2ps Compare the natural frequency of the device described in Problem 10.1ps when it is on the Earth to its natural frequency when it is on the Moon where the acceleration due to gravity is 5.31 ft/s^2.

10.3ps The figure depicts a 200-lb bungee jumper. The cord has an unstretched length of 60 ft and stretches an additional 40 ft before the jumper stops. The jumper asks you to determine whether the same cord could be used to make a jump on the Moon where the acceleration due to gravity is 5.31 ft/s^2. He would jump from a tower 90 ft above the lunar surface. He would be wearing 150 lb (Earth weight) of life support gear. In his life support gear he is 6 ft tall and the cord is tied to his feet. Determine the distance between the top of his head and the surface when he comes to a stop. If you model the bungee cord as a linear spring, what are the periods of his vertical oscillations on Earth and on the Moon?

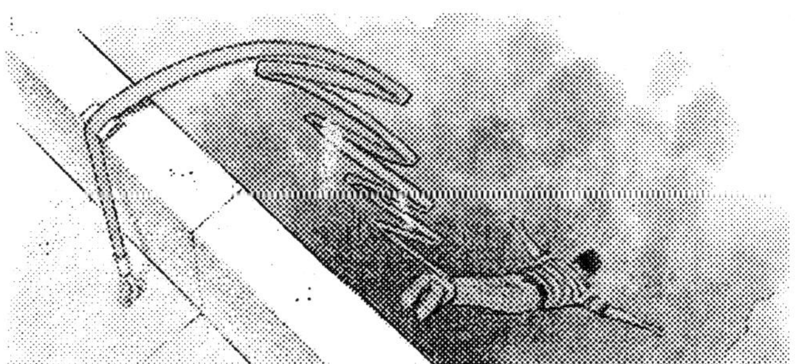

10.4ps Problem 10.3ps compares bungee jumping on the Earth and on the Moon. Find the periods of the jumper's vertical oscillations on Earth and on the Moon.

10.5ps The pendulum consists of a slender bar rigidly attached to a homogeneous 1-kg disk. Let L be the bar's length (which is 60 mm in the figure), and let the mass of the bar be 1/3 kg per meter. Draw a graph of the natural frequency of small vibrations of the pendulum as a function of L for $0 \leq L \leq 1000$ mm.

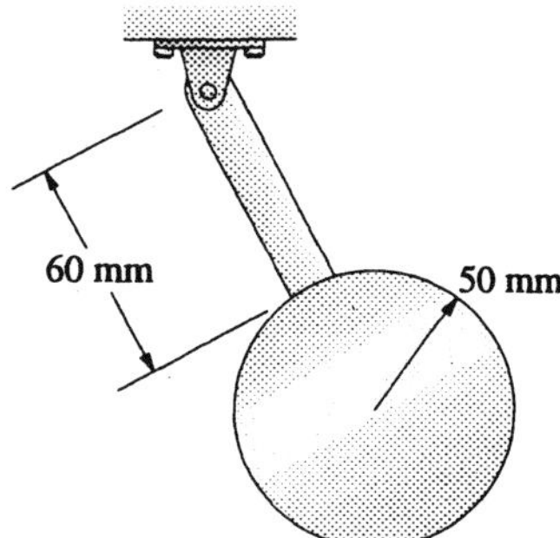

10.6ps The mass of the oscillator is 3 kg, $k = 12$ N/m and $c = 12$ N-s/m. The position x is measured from the position of the mass when the spring is unstretched. At $t = 0$ the mass is released from rest at $x = 0.1$ m. Determine position of the mass as a function of time.

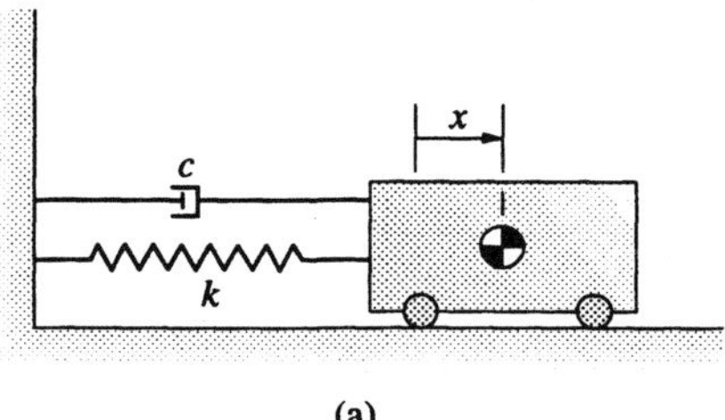

(a)

10.7ps In Problem 10.6ps, suppose that the damping constant is increased to $c = 20$ N-m/s. Determine the position of the mass as a function of time.

Statics Supplement
Answers to Even Numbered Problems

1.2ps Length = 63.7 m, span = 60.9 m, Height = 18.4 m, diam = 5.87 m

1.4ps Altitude = 60,700 ft

1.6ps Weight = 964,052 lb

1.8ps Distance = 24,508.08 miles. Speed = 115.646 mph = Speed = 186.114 km/hr = 169.614 ft/s = 100.493 knots.

2.2ps | **F** | = 16.6 kN

2.4ps Length OAB = 1.294 km, cost OAB = \$647,000 Length OB = 1.200 km, cost OB = \$720,000

2.6ps For $\mathbf{r}_A$, $\cos\theta_x = 0.447$,$\cos\theta_y$ 0.894, $\cos\theta_z = 0$
For $\mathbf{r}_B$, $\cos\theta_x = 0.772$, $\cos\theta_y = -0.154$, $\cos\theta_z = -0.617$.
Distance = 0.574 km

2.8ps

$$\cos\theta_x = (10 + b)/ (552 + 20\,b+b^2)^{1/2}$$

$$\cos\theta_y = -\,16 / (552 + 20\,b + b^2)^{1/2}$$

$$\cos\theta_z = -\;14 / (552 + 20\,b + b^2)^{1/2}$$

2.10ps Distance = 5.44 ft

2.12ps Distance = 6481 meters

2.14ps $\mathbf{r}_{CA} \times \mathbf{F} = 102.9\,\mathbf{i} + 61.74\,\mathbf{j} - 77.17\,\mathbf{k}$ (N-m)

2.16ps Component = 6.19 kN

2.18ps $\mathbf{r}_{AD} \times \mathbf{F}_B = -2.41\,\mathbf{i} + 1.87\,\mathbf{j} + 4.01\,\mathbf{k}$ (kN-m)

3.2ps $A_x = 600$ N
B = 2040 N

3.4ps Keep $\alpha \leq 14.2$

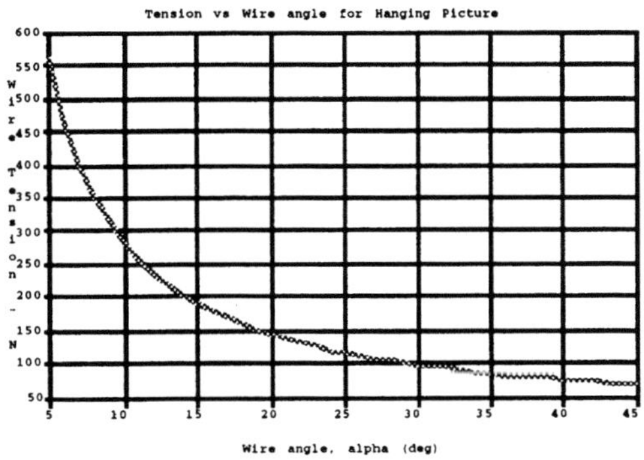

3.6ps $W_{max} = 1414$ lb

3.8ps $\Delta h = 10.28$ m

3.10ps

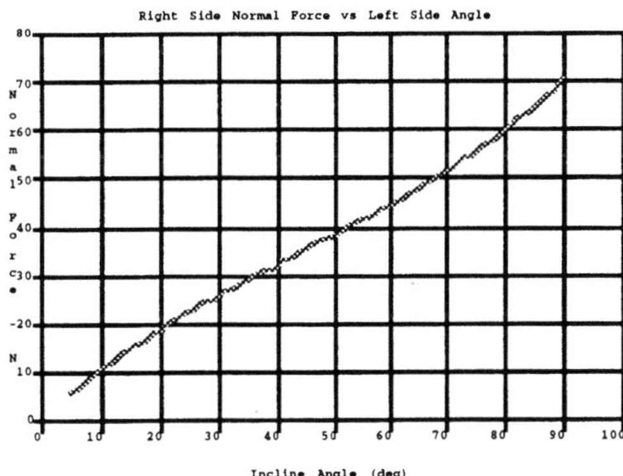

The right side normal force is smallest where the left side angle is smallest. The reason for this is that the horizontal component of the left side normal force is smallest and the vertical component on the left side is largest for the smallest left side angle. If the left side angle were zero, the right hand normal force would be zero as the left side normal force would completely support the cylinder.

3.12ps $x_A = -\,y\,b / h$
$y_A = (h - b\tan\theta) / 2$

3.14ps Horizontal force = 162.7 N

4.2ps d = 0.583 m

4.4ps The moment of the force F about point P for the angle a varying over the range $0 \le a \le 90°$ is shown below.

Moment (N - m) vs Angle (deg)

Moment - N m

Angle (deg)

4.6ps

Sum of Moments of F1 and F2 about A vs xp

Moment - Sum - N m

Distance, xp (m) , from A to Application of F2

4.8ps Fequiv = 2.91 **i** - 333.5 **j** lb
distance = 5.27 ft

4.10ps M_A = -698 **k** ft lb

4.12ps P is located at (3,-2,0)

4.14ps The moment of the couple is plotted as the angle a varies over the range0° ≤ a ≤ 180°.

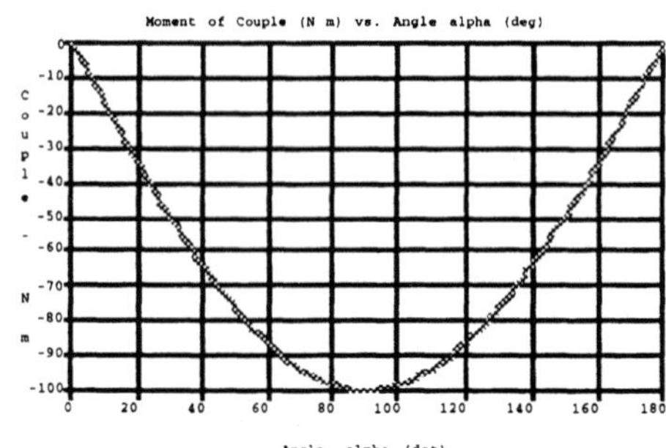

4.16ps **F** = 100 **j** + 80 **k** N
M = 120 **j** - 150 **k** N-m

4.18ps **F** = 28 **k** kip
M = 96 **i** + 192 **k** ft-kip

4.20ps x = - 0.714 ft
z = - 0.171 ft

4.22ps A_x = -546.4 N
A_y = -98.4 N
A_z = 48.4 N

4.24ps **F** = 70 **j** N
x = 0.771 m

5.2ps F = 118.2 lb

5.4ps (a) V = 0
F_R = 80 l
M = - 134 ft-lb
(b) V = 0
F_R = 80 lb
M = -159.3 ft-lb

Longer arms require larger internal moments at the shoulder.

5.6ps There is change in the support reactions.

5.8ps F = W/2

5.10ps

Force (kip) in line AB vs Boom Angle Theta (deg)

FAB - kip

Boom Angle - Theta (deg)

5.12ps (b) A_x = 3.69 kN
A_y = 22.1 kN
A_z = 11.1 kN
M_{Ax} = 1 .77 kN-m
M_{Ay} = -2.06 kN-m
M_{Az} = 3.54 kN-m

5.14ps

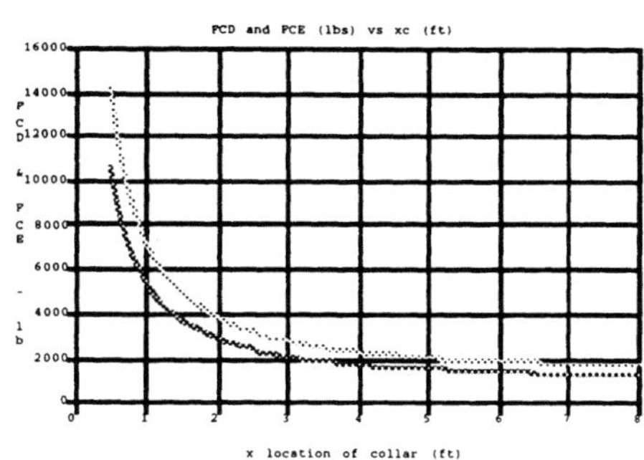

5.16ps

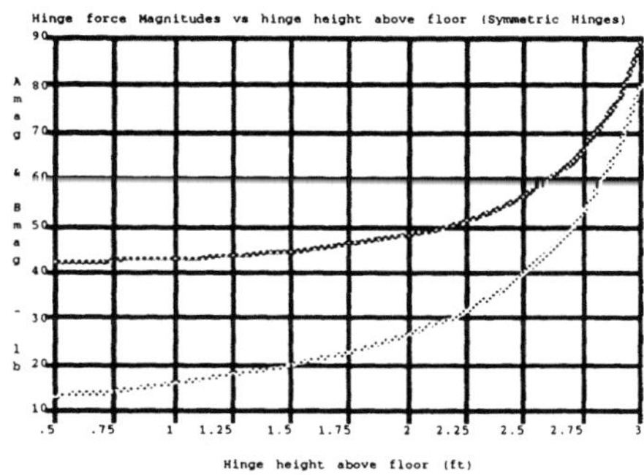

5.18ps

Force Components Ax, Ay, and Bx (lb) vs the distance AB (ft)

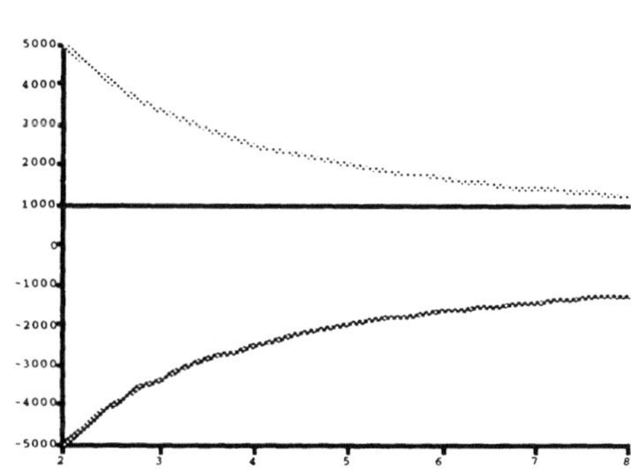

5.20ps

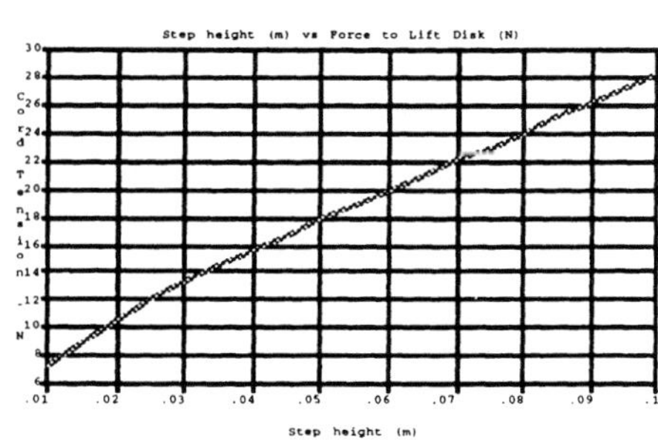

5.22ps

$A_x = -57.1$ lb

$A_y = 85.7$ lb

$T_{BC} = 127.8$ lb

5.24ps

$T = 16.42$ N

5.26ps

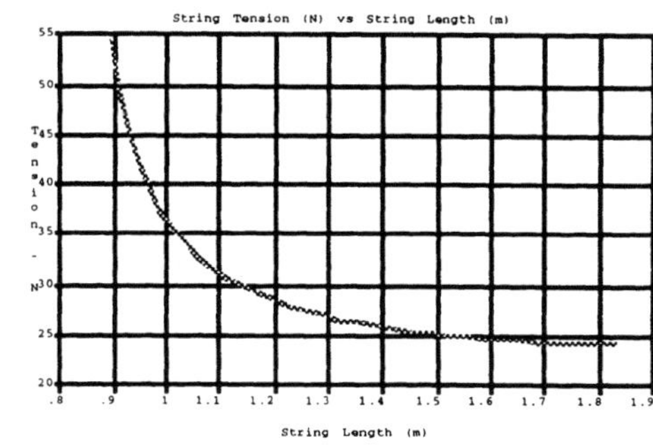

5.28ps

(b) $A_y = 700$ N

$B_x = 0$ N

$B_y = -600$ N

5.30ps

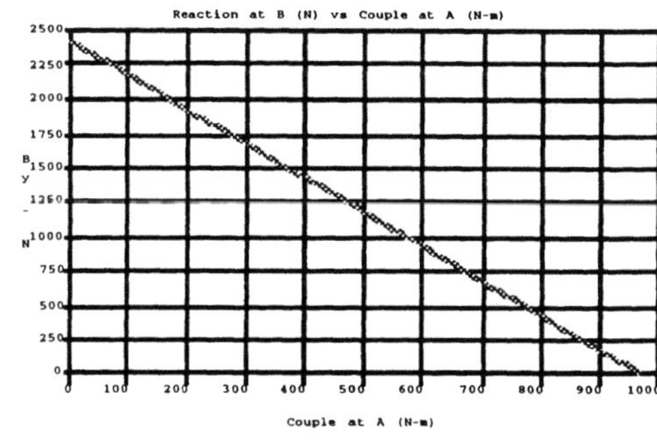

(a) $C_{max} = 971.2$ N-m

(b) See Plot

5.32ps

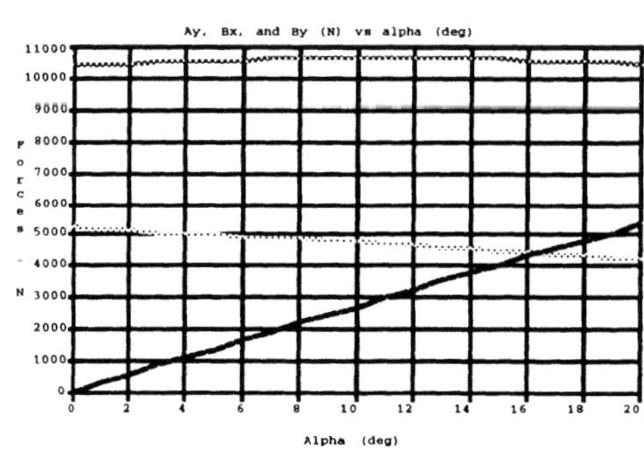

At the left hand axis, the order of thecurves, bottom to top, is Bx, By, Ay.

Bx and By cross at about 16 degrees. At this point, the friction force required for the car not to slip must be equal to the normal force on the front wheels. As we will see later, this will not happen. In fact, slipping will occur where the friction force is some fraction, ms, times the normal force at that wheel. Thus, equilibrium will be limited to angles of incline somewhere below 16 degrees if we do not lock the rear wheels. Note that Ay is about twice as large as By it might be a good idea to put brakes on the

5.34ps $A_x = -25.4$ kN

$A_y = -34.3$ kN

$F_{BC} = 68.6$ kN

5.36ps $F_C = 0.5$ W

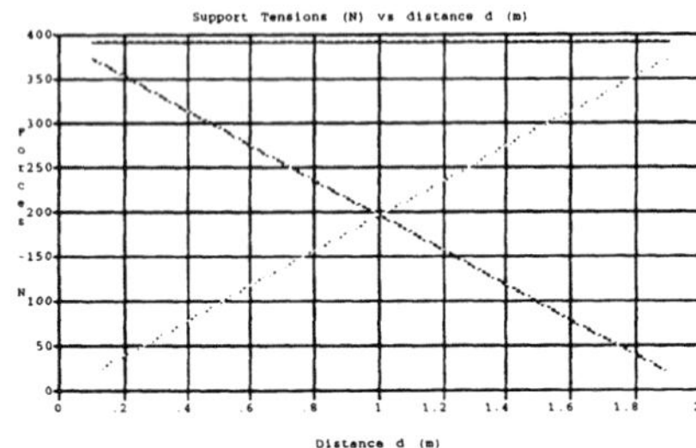

FC is constant, FB is small when d is small, and FA is small when d is large. When C is at the same end of the plate as A or B, both cords support that end of the plate and do not have to carry as large a load as when one cord supports an end of the plate alone.

5.38ps $A_x = 22.3$ lb

$A_y = 80$ lb

$A_z = -33.4$ lb

$F_{BC} = 26.0$ lb

$N_z = 20.1$ lb

6.2ps (a) $A_x = 0$

$A_y = 19\ F/4$

$F_y = 21\ F/4$

(b)

AB	$= 15\ F/4$	(T)
AG	$= -15\ F\ \sqrt{2}/4$	(C)
BC	$= 15\ F/4$	(T)
BG	$= 2\ F$	(T)
GC	$= 7\ F\ \sqrt{2}/4$	(T)
GH	$= -11\ F/2$	(C)
HC	$= 0$	
HI	$= -11\ F/2$	(C)
CI	$= 5\ F\ \sqrt{2}/4$	(T)
CD	$= 17\ F/4$	(T)
IF	$= -17\ F\ \sqrt{2}/4$	(C)
ID	$= 3\ F$	(T)
DF	$= 17\ F/4$	(T)

6.4ps (a) $T_{AC} = -507.7$ lb (C)

$T_{AB} = 954.2$ lb (T)

(b) $T_{AC} = -780.2$ lb (C)

$T_{AB} = 1036.8$ lb (T)

(c) $T_{AC} = -1130.5$ lb (C)

$T_{AB} = 1226.7$ lb (T)

6.6ps (a) $F_{max} = 290$ N

(b) $T_{BC} = 301.6$ N (T)

$T_{BA} = -82.9$ N (C)

$T_{CA} = 46.1$ N (T)

6.8ps $T_{AB} = -642.8$ lb (C)

$T_{AC} = 214.3$ lb (T)

$T_{CD} = 214.3$ lb (T)

$T_{BC} = 600$ lb (T)

$T_{BD} = -303.0$ lb (C)

6.10ps AC = 4 ft mimizes T_{BC} at a value of 600 lb

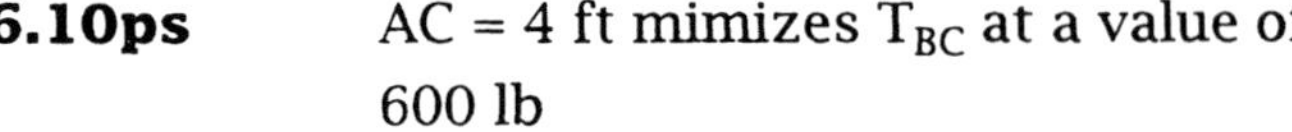

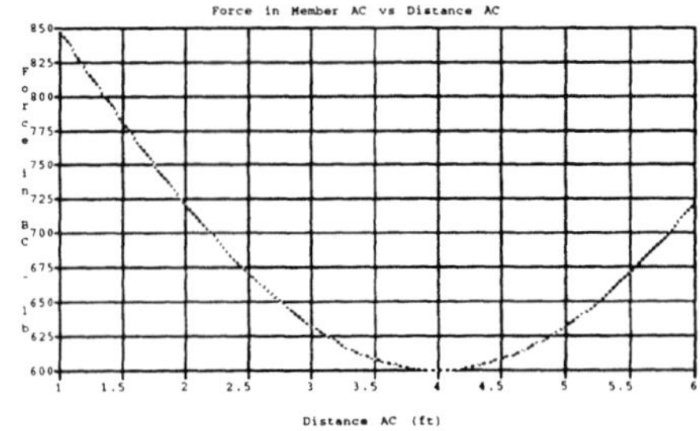

6.12ps You can lower the loads in some members, but the constant compressive loads in members AC and CD limit the load F regardless of the force G.

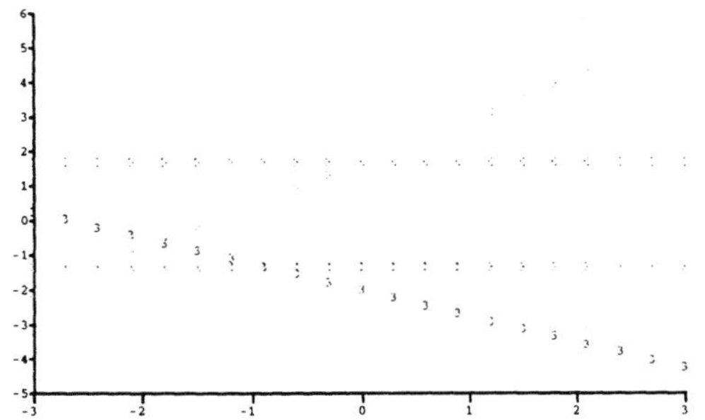

6.14ps The height BC = 8 ft mimizes the load in BC. The load when BC = 8 ft is11.2 kN.

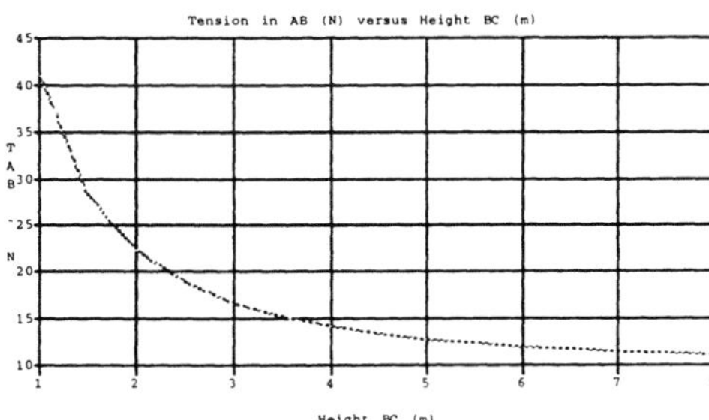

Obviously, the minimum load in AB occurs when BC = 8 m. The value for the load in AB when BC = 8 m is 11.2 kN.

6.16ps $T_{AB} = 0$ when F_1 = - 20,000 lb and F_2 = 40,000 lb

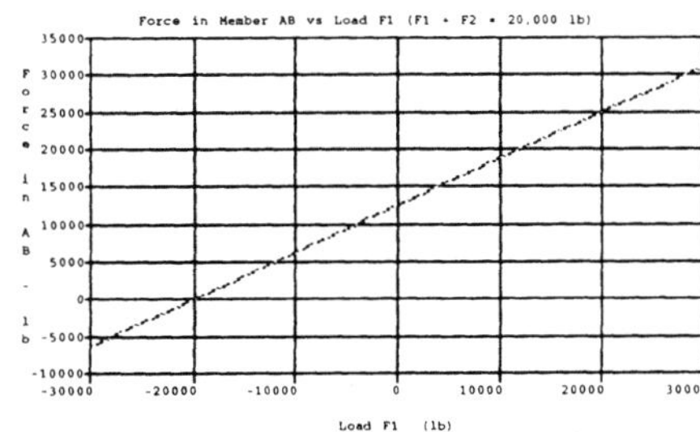

6.18ps B = 6000 N upward causes the minimum load magnitudes in the other members.

Plot 1 shows the forces in all members as a function of the force B.

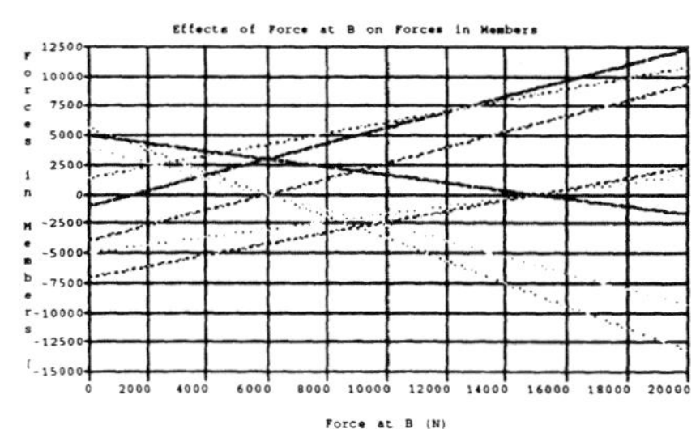

The second plot zeros in on the range B – 4000 lb to 8000 lb.

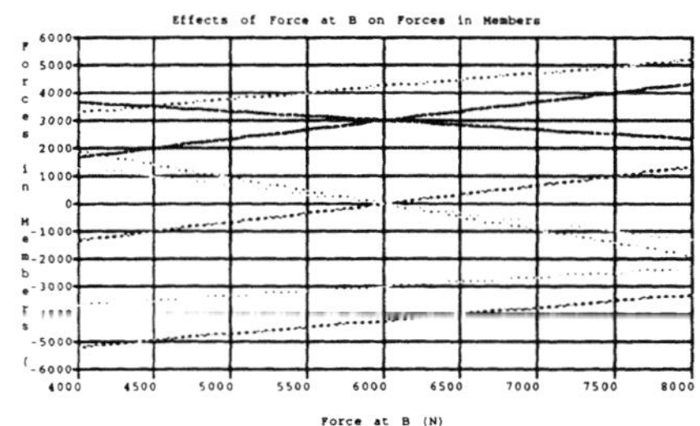

By inspection, the minimum occurs near B = 6000 N. At this value, the minimum loading occurs for the maximum loaded element.

6.20ps

$C_x = 1.2$ W
$C_y = 3.5$ W
$T_{AB} = 1.92$ W (T)
$T_{BD} = 0.64$ W (T)
$T_{BE} = 8$ W (T)
$T_{BC} = -2$ W (C)
$T_{CD} = -1.92$ W (C)
$T_{DE} = -1.28$ W (C)

6.22ps $m_{max} = 72.08$ kg

6.24ps Design problem - no unique solution

6.26ps T_{CJ} = 141.4 kN in both solutions. T_{CD} and T_{IJ} are different in the two solutions.

6.28ps

(a) $F_y = 510$ N
$B_x = 0$
$B_y = 120$ N

(b) CE = 240 N (T)
BE = 100 N (T)

BD = -320 N (C)
GF = -150 N (C)
GH = 400 N (T)
HF = - 500 N (C)

6.30ps F_{CG} = 3000 lb (T)

6.32ps F_H = -11 F/3 to minimize load in member JK (F_{JK} = 0)

6.34ps (b) F_{CJ} = - 2.5 kN (C) with no wind
F_{CJ} = - 3.33 kN (C) with wind

6.36ps Multiple solutions possible. One possible solution is F_F = 20 kip, $F_A = F_B = F_D$ = 0. This gives F_{GI} = -6/67 kip (C).

6.38ps

Original Truss	Redesigned Truss
HI =	HI =
GI =	GI =
KI =	KI = 0

6.40ps (a) F_{AB} = -376 lb (C)
F_{AC} = -594 lb (C)
F_{AD} = 819 lb (T)

6.42ps

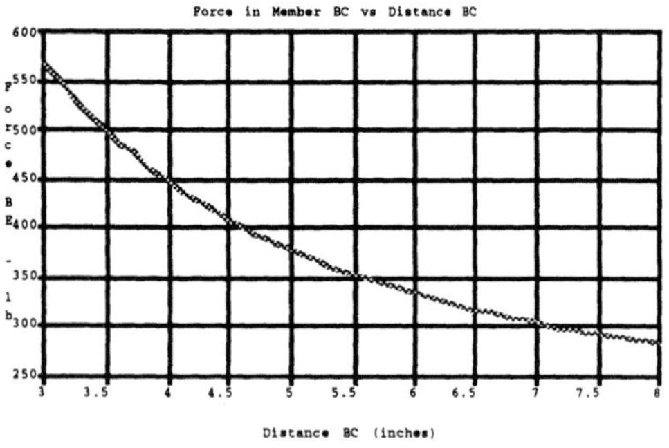

Force in Member BC vs Distance BC

Force BC (lb): 250, 300, 350, 400, 450, 500, 550, 600

Distance BC (inches): 3, 3.5, 4, 4.5, 5, 5.5, 6, 6.5, 7, 7.5, 8

6.44ps (a) $B_x = 0$
B_y = - 350 N
C_y = 150 N
(b) $A_x = 0$
A_y = 500 N
M_A = 500 N-m

6.46ps Support reactions
Ax = - 200 N
Ay = 450 N
Cy = 350 N
Forces between frame members
ax = 100 N acting on AC
ay = -450 N acting on AC
cx = 100 N acting on AC
cy = -350 N acting on AC
Bx = -100 N acting on AB
By = -450 N acting on AB

6.48ps Ax = 14.4 lb
Ay = -39.7 lb
Cx = -50.4 lb
Cy = -24.3 lb

7.2ps x = (2 + 2/n)/(3 + 6/n)
y = (2 + 2/n)/(6 + 3/n)
As n becomes large, the area approaches that of the right triangle formed by the x axis, the line x = 1, and the line AB.

7.4ps x —> 3/4 as n becomes large

7.6ps Centroid is at (0.373, 1.25)

7.8ps $x_C = [2b^2R + 2\ c\ R\ (b+c/2) + \{b+c+(4R)/(3p)\}(pR^2)/2]\ /\ A_{total} - [(b+c/2)((pR^2)/4]/\ A_{total}$

$y_C = [\ (\ 2R^2b/3 + R(2cR + (pR^2)/2 - (pR^2)/4]\ /\ A_{total}$

$A_{total} = R(b + 2c) + (pR^2)/4$

7.10ps Distance = 1.83 mm

7.12ps $x_{centroid}$ = 131.7 mm,
$y_{centroid} = z_{centroid} = 0$

7.14ps $x_{centroid}$ = 942.6 mm
$y_{centroid}$ = 899.1 mm

8.2ps $x_{centroid}$ = 1 ft,
$y_{centroid}$ = 1.5ft
$I_{x'} = 8.5\ ft^4$
$I_{y'} = 4\ ft^4$

8.4ps $I_{yaxis} = 7.404 \times 10^{-5}\ m^4$

8.6ps Ix'y' = 0 by symmetry

8.8ps IPtotal = (2 /3)m L2

8.10ps Ixcircle = 1.89 x 10-6 m4
Ixsquare = 1.18 x 10-6 m4

8.12ps Ixcm = 9.37 x 10-3 slug ft2
Iycm = 17.41 x 10-3 slug ft2
Izcm = 26.79 x 10-3 slug ft2
xcm = 0.444 ft , ycm = 0.295 ft

$k_y = 87.82$ mm

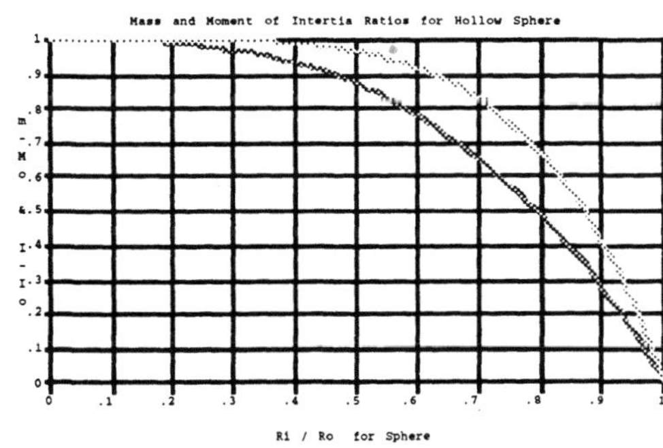

8.14ps Top curve is I / Io Ratio

9.2ps Load = 5760 lb at x = 9 ft.

9.4ps Multiple answers are possible. Study the example problem care fully and follow it.

9.6ps Bottom curve is m / Mo Ratio

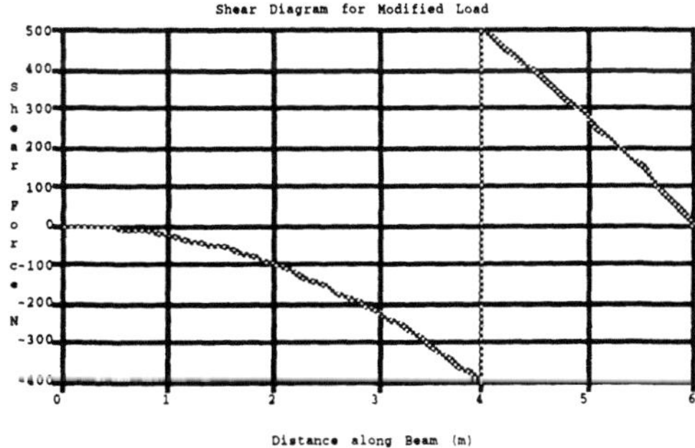

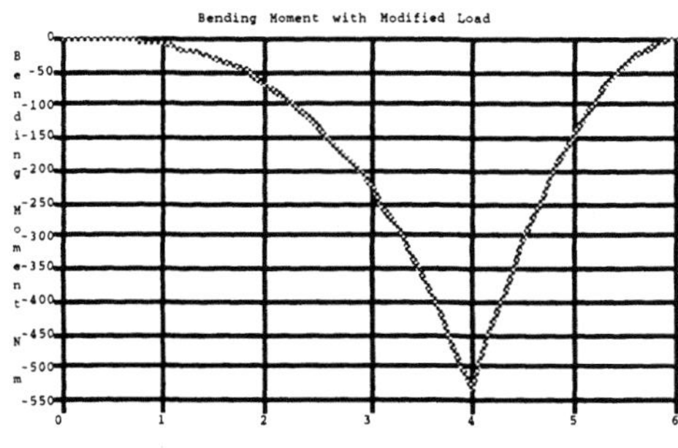

10.2ps Design problem -no unique answer. Hint: consider constant speed sliding.

10.4ps N/A

10.6ps h < b / ms and F < ms NB

10.8ps N1 = 423.1 lb
N2 = 1142 lb
N3 = 1085 lb
F = 640 lb

10.10ps a = 16.7°

10.12ps

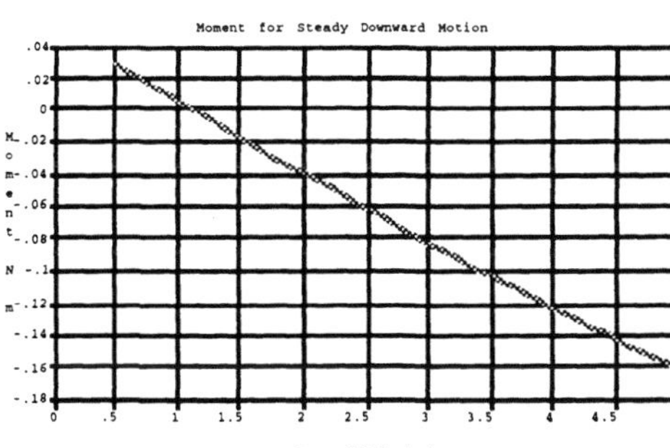

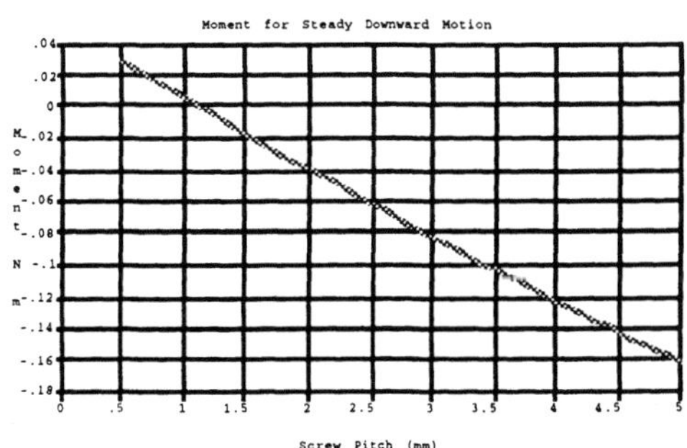

10.14ps

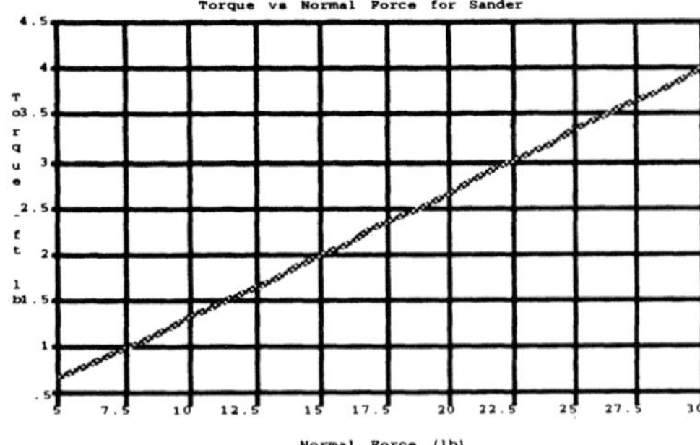
Torque vs Normal Force for Sander
Torque_ft lb
Normal Force (lb)
4.5
4
3.5
3
2.5
2
1.5
1
.5
5
7.5
10
12.5
15
17.5
20
22.5
25
27.5
30

DynamicsSupplement
Answers to Even Numbered Problems

1.2ps $W_E = 152.5$ lb

1.4ps v = 11,567 m/s

1.6ps Volume = 801.3 ft^3
Cube side = 9.29 ft

1.8ps v = 98.35 m/s
v = 322.7 ft/s
v = 354.1 km/hr

1.10ps c = 186,282.4 mi/s
c = 6.606167 x 10^8 mi/hr
c = 1.079253 x 10^9 km/hr

1.12ps

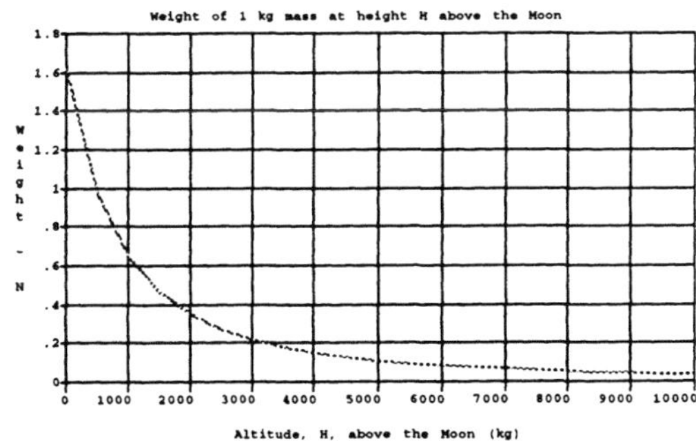

2.2ps (a) Displacement = 1840 ft

(b) v (3) = 140 ft/s

a (2) = 64 ft/s^2

2.4ps $y = 0.9\ g\ (\ t + e^{-ct}/c)\ /c$

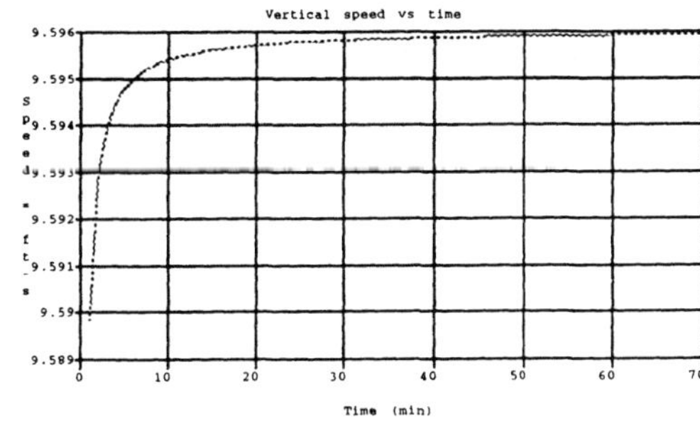

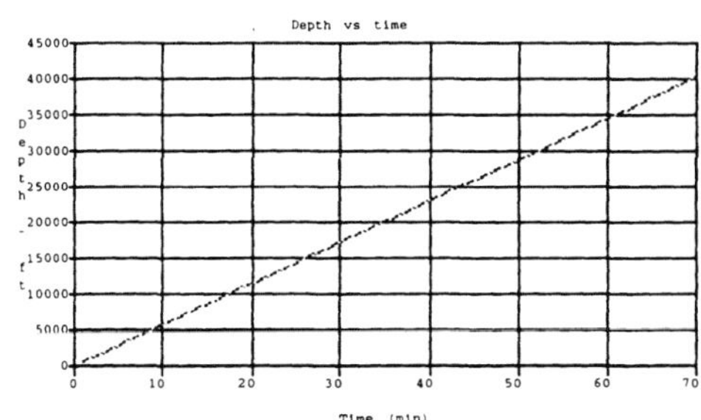

2.6ps $d = x_{impact} = 447$ m

2.8 ps $d_{miss} = 14$ m

2.10ps

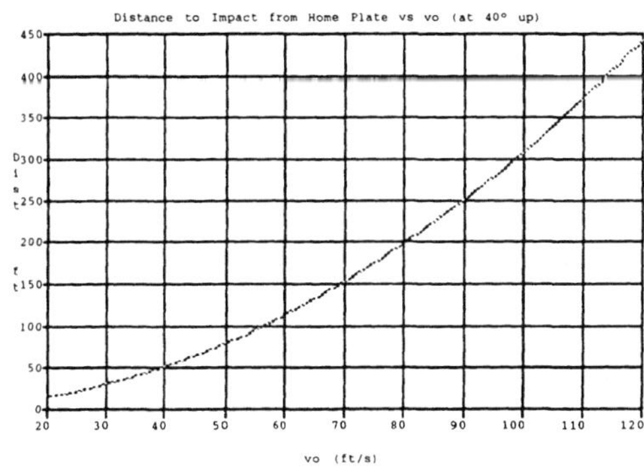

2.12ps C = 164.5 rad/s^2

2.14ps $\omega_{Mars} = 7.095 \times 10^{-3}$ rad/s
$\omega_{Mars} = 4.065 \times 10^{-3}$ deg/s

2.16ps $\omega = 26.39$ rad/s

2.18ps

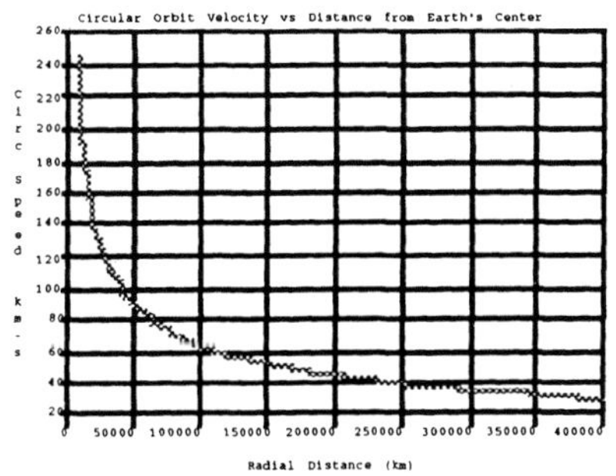

2.20ps a(80 ft) = 39.88 ft/s^2
a(midpoint) = 7.82 ft/s^2
a(B - 80 ft) = 65.40 ft/s^2

2.22ps

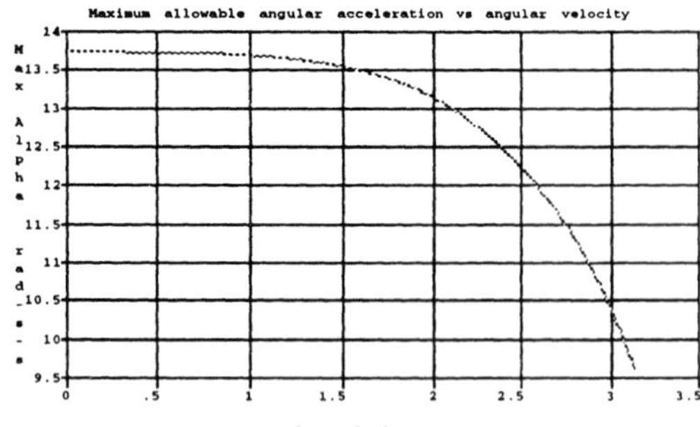

2.24ps $h_{max} = 28,020$ ft
$v(\text{at } h_{max}) = 232.4$ ft/s

2.26ps $\mathbf{v} = 282\ \mathbf{u}_r - 528\ \mathbf{u}_\theta$ m/s
$\mathbf{a} = -1680\ \mathbf{u}_r - 936\ \mathbf{u}_\theta\ m/s^2$

2.28ps $a_T = 34.16\ m/s^2$, $a_N = 1932\ m/s^2$

2.30ps $|a| = 172.2\ ft/s^2 = 5.35$ g

2.32ps $v_p = 10.11$ km/s, $v_a = 1.68$ km/s

2.34ps $v_c = 1632$ m/s,
Period = 7078 seconds (117 min, 58 s)

2.36ps $v_a = 1,586$ mph, $\phi = 69.5°$,
Heading = 20.5° East of North.

2.38ps $\mathbf{r} = 40.9\ \mathbf{u}_r$ m

$\mathbf{v} = 40.8\ \mathbf{u}_r + 215.9\ \mathbf{u}_\theta$ m/s

$\mathbf{a} = -1104\ \mathbf{u}_r + 544.0\ \mathbf{u}_\theta\ m/s^2$

3.2ps At t = 0.5 s, **F** = 4.66 **i** + 45.98 **k** lb

At t = 2 s, **F** = 4.66 **i** lb

3.4ps Scales read zero

3.6ps $a = 15.3\ ft/s^2$

3.8ps Miss = 1.86 ft

3.10ps s = 0.44 ft

3.12ps Ratio = 2.46

3.14ps $a_{max} = 329.7\ ft/s^2 = 10.24$ g

3.16ps Track banking is designed for zero side force

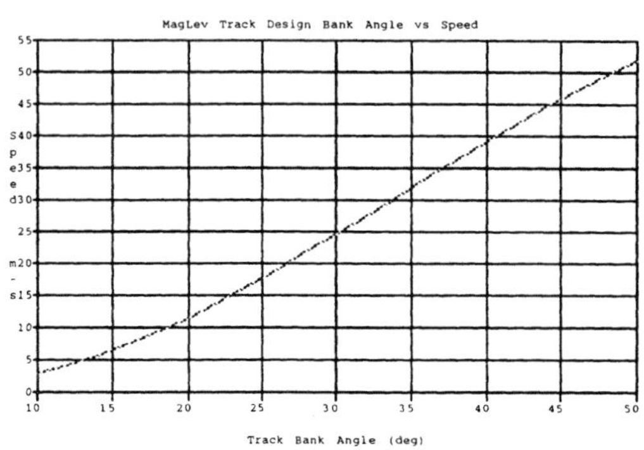

3.18ps Maximum Turn Speeds vs Turn Radius for μs = 0.1, 0.3, 0.5, and 0.7

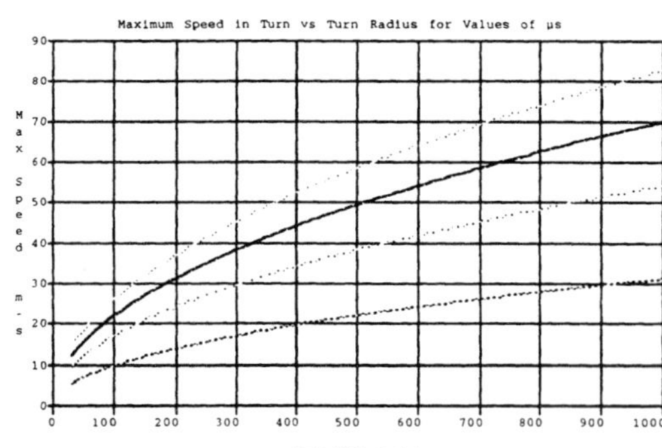

3.20ps

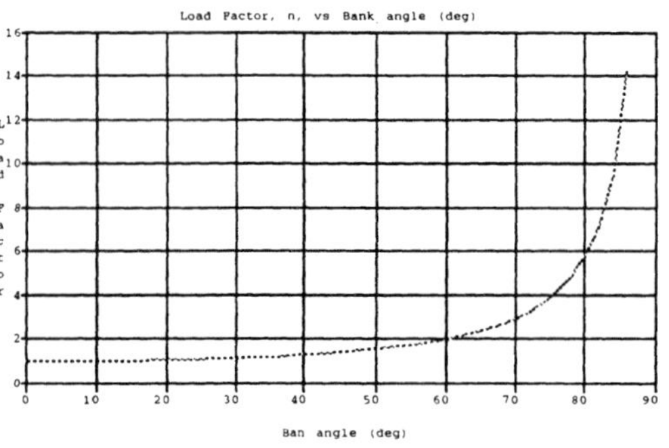

3.22ps $v_o^2 = 2\, m\, g\, x$

Bottom curve is with anti-skid brakes (μs = 0.6). Top curve is with wheels locked (μs = 0.4)

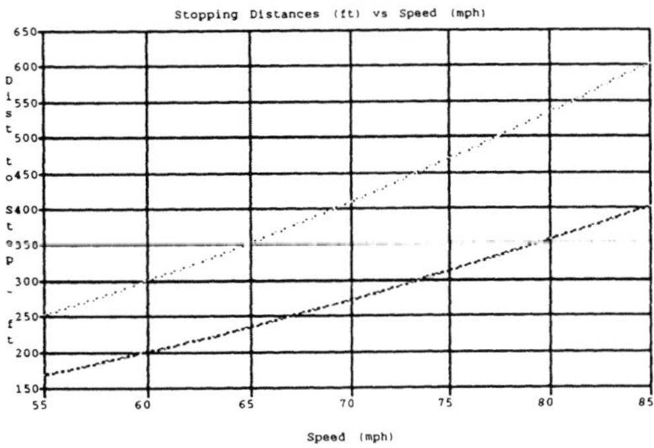

3.24ps

Car barely maintains contact with road at the top ofthe hill

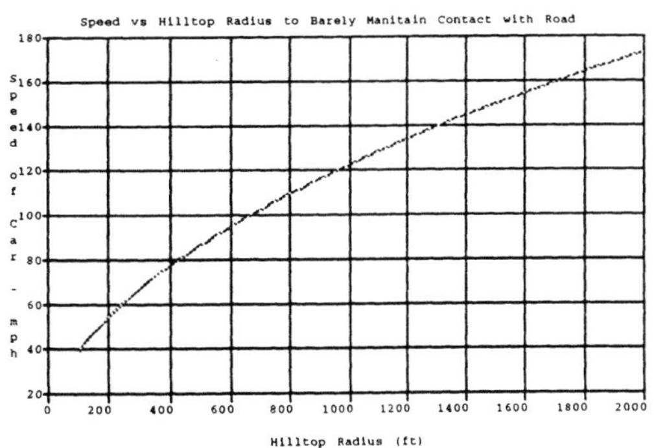

3.26ps $F_r = 6.74$ N

$F_q = -2.17$ N

3.28p Load factor = n = 5.50 g

3.30ps Note, when the load factor is nega tive, the passengers must be strapped into the car and the car must be held on the track by some mechanism.

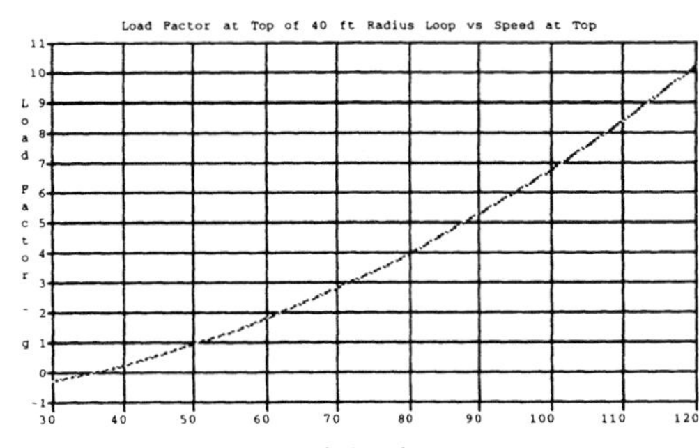

4.2ps xf = 1,010 ft

4.4ps Without rocket assist X_{TO} = 518 ft

Rocket Thrust to halve

X_{TO} = 25,440 lb

4.6ps Both ratios = 2.44

4.8ps

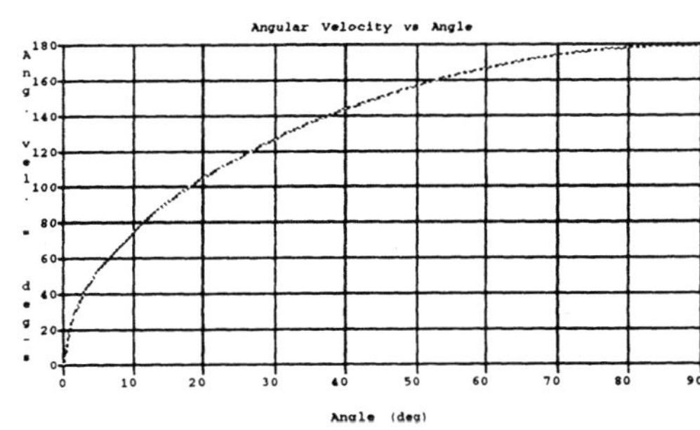

4.10ps (a) $v_E = 3.25$ m/s

(b) $v_M = 1.32$ m/s

(c) Ratio = 0.406

4.12ps $(5\, g\, R\,)^{0.5} \leq v_o \leq (6.5\, g\, R)^{0.5}$

4.14ps Let a_1 be the angle at which contact is lost.

$\alpha_1 = 48.19°$ and is independent of the value of g

4.16ps $h_f = 1365$ ft

4.18ps Power = 1,610 hp

4.20ps Work = - 1 N m

4.22ps $_\theta = 37.54°$

b = 6.21 ft

4.24ps Earth, Distance = 6 ft
Moon, Distance = 0.99 ft

5.2ps $\mathbf{v}(5s) = 12.5\,\mathbf{i} + 51\,\mathbf{j} - 62\,\mathbf{k}$ m/s

5.4ps $v_f = 75.7$ ft/s

5.6ps $t_{stop} = 0.982$ s
$x_{stop} = 0.218$ m

5.8ps L.I. = 150 N s

5.10ps

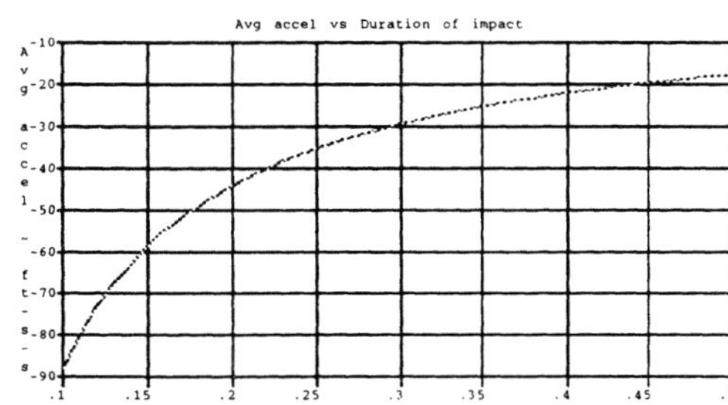

5.12ps $F_{avg} = -85.2$ lb
L.I. = - 3.41 lb s

5.14ps $|a_{avg}| = 21200$ ft/s^2 = 657 g

5.16ps Distance = 2.00 ft

5.18ps Car A is the bottom line and Car B is the top line.

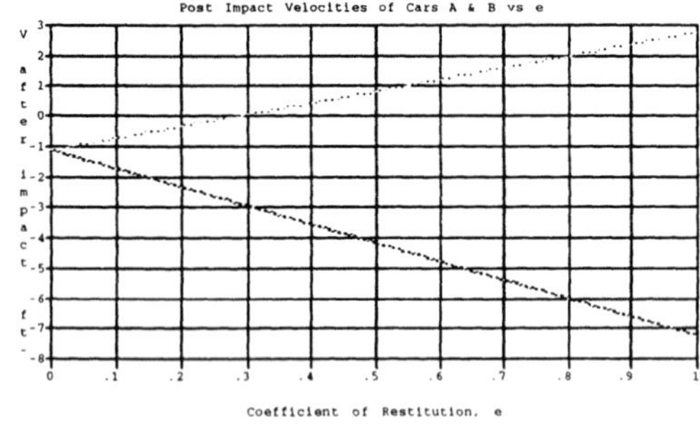

5.20ps $v_a = 94.5$ m/s, $v_p = 10{,}870$ m/s

5.22ps $\mathbf{h} = -80\,\mathbf{i} - 10\,\mathbf{j} + 150\,\mathbf{k}$ kg m^2/s

5.24ps T = 648 lb

5.26ps v = 5450 m/s

5.28ps (a) $v_a = 1.67$ km/s
(b) $v_{transverse} = 2.5$ km/s

5.30ps (a) **L.I.** = $15.4\,\mathbf{i} + 12.9\,\mathbf{j}$ lb s
(b) $x_{impact} = 311.7$ ft

5.32ps With both improvements, $V_{max} = 85.7$ ft/s

6.2ps $|\mathbf{a}| = 171.3$ ft/s^2

6.4ps Angular velocity, $\omega = 40$ rad/s clockwise
Angular acceleration, $\alpha = 40$ rad/s clockwise

6.6ps

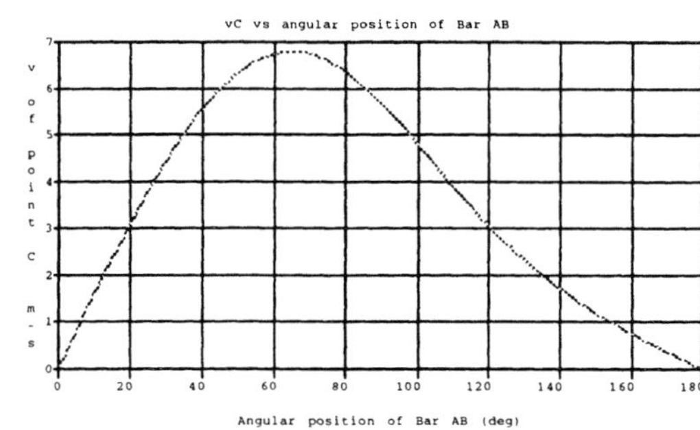

6.8ps (a) $\omega = 30\,\mathbf{i}$ rad/s
(b) $\omega = -30\,\mathbf{i}$ rad/s

6.10ps (a) Disk is slipping
(b) $\mathbf{v}_B = -9\,\mathbf{i}$ ft/s
$\mathbf{v}_C = 3\,\mathbf{i} + 12\,\mathbf{j}$ ft/s
$\mathbf{v}_D = 11.49\,\mathbf{i} + 8.49\,\mathbf{j}$ ft/s

6.12ps $\mathbf{v}_{midpoint} = -20\,\mathbf{i} - 10\,\mathbf{j}$ ft/s

6.14ps $\omega_{BC} = 5.33\,\mathbf{k}$ rad/s (counterclock wise)
$\omega_{CD} = 5.33\,\mathbf{k}$ rad/s (counterclock wise)

6.16ps $\omega_{BC} = 5.33\,\mathbf{k}$ rad/s
$\mathbf{v}_C = 2.29\,\mathbf{i} + 0\,\mathbf{j}$ m/s

6.18ps $\mathbf{v}_B = -1.2\,\mathbf{i} + 0.60\,\mathbf{j}$ m/s
$\mathbf{v}_C = -0.665\,\mathbf{i} - 0.470\,\mathbf{j}$ m/s
$\mathbf{v}_D = -1.38\,\mathbf{i} - 0.83\,\mathbf{j}$ m/s

6.20ps v_{max} to left = -3.35 ft/s
v_{max} to right = 3.75 ft/s

Saw blade speeds — Negative speeds to left, Positive speeds to right.

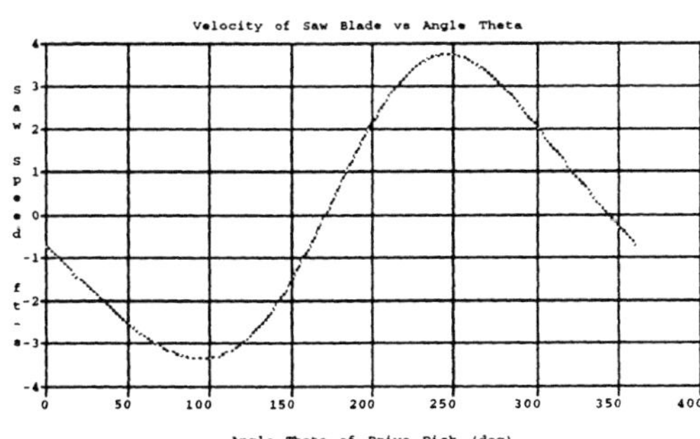

The speeds around the maximum and minimum speeds are tabluated below:

Angle (deg)	Speed of C (ft/s)
93	-3.3433464
94	-3.3445767
95	-3.3447271 *Max Speed to Left*
96	-3.3437824
97	-3.3417277
244	3.75055424
245	3.75280377
246	3.75344895*Max Speed to Right*
247	3.75250059
248	3.74997078

6.22ps $\omega_{BC} = -5\ \mathbf{k}$ rad/s
$\omega_{CD} = -3.54\ \mathbf{k}$ rad/s

6.24ps (a)Instantaneous center is at (1.15, 2) m
(b)$\mathbf{v}_B = -1.5\ \mathbf{i} + 2.60\ \mathbf{j}$ m/s

6.26ps $\omega_{BC} = 2.4\ \mathbf{k}$ rad/s (counterclock wise)

6.28ps (a)Topmost point, Q, on wheel has greatest velocity magnitude
$|\mathbf{v}_Q| = 4$ m/s
(b)Q also has greatest acceleration magnitude
$|\mathbf{a}_Q| = 7.21$ m/s^2

6.30ps a = 2.72 ft/s2 to the left

6.32ps

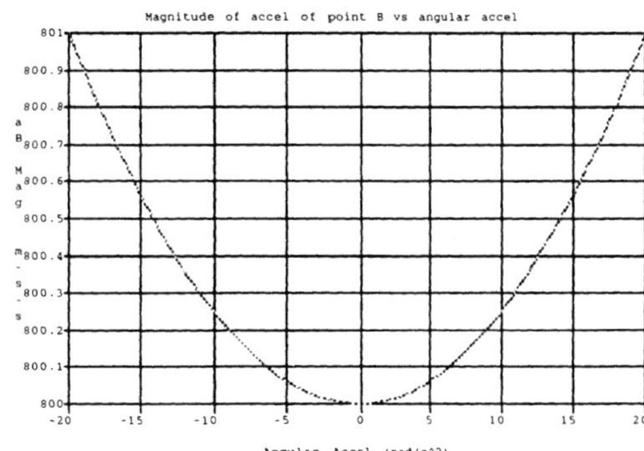

6.34ps $\mathbf{v}_B = -1.5\ \mathbf{i} + 2.60\ \mathbf{j}$ m/s
$\mathbf{a}_B = 3.10\ \mathbf{i} - 5.37\ \mathbf{j}$ m/s^2

6.36ps $\mathbf{v}_Q = 2.12\ \mathbf{i} - 2.12\ \mathbf{j}$ ft/s
$\mathbf{a}_Q = 17.45\ \mathbf{i} - 12.9\ \mathbf{j}$ ft/s^2

6.38ps $\mathbf{a}_C = -244\ \mathbf{i}$ m/s^2

6.40ps $\mathbf{v}_C = 3.13\ \mathbf{i}$ ft/s
$\mathbf{a}_C = -14.72\ \mathbf{i}$ ft/s^2

6.42ps

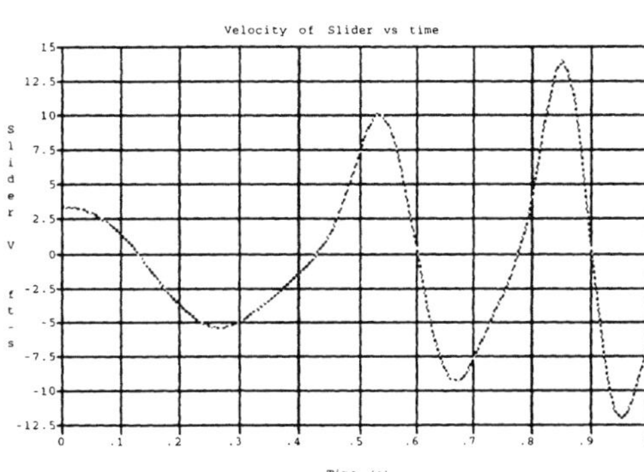

6.44ps

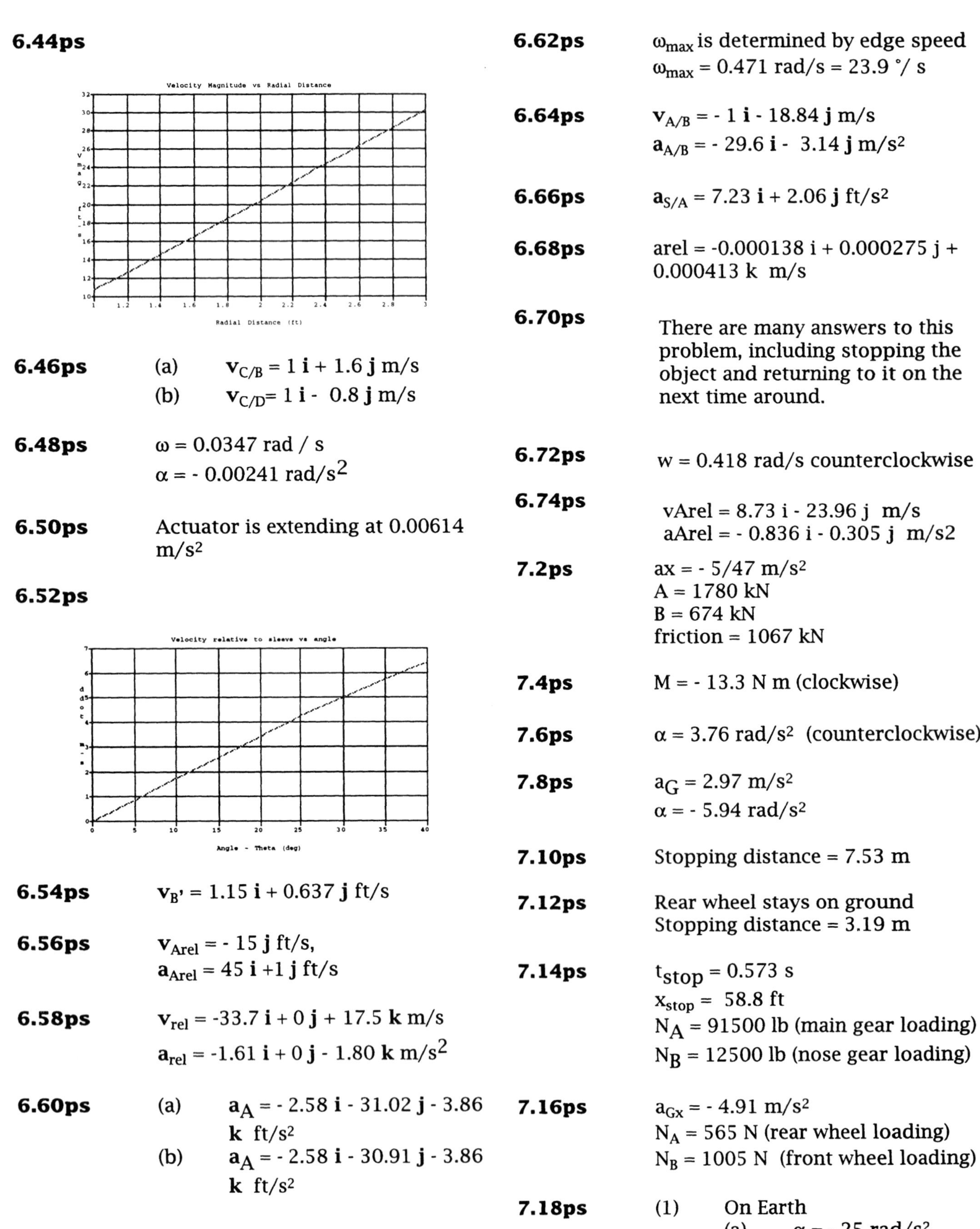

6.46ps (a) $\mathbf{v}_{C/B} = 1\,\mathbf{i} + 1.6\,\mathbf{j}$ m/s
(b) $\mathbf{v}_{C/D} = 1\,\mathbf{i} - 0.8\,\mathbf{j}$ m/s

6.48ps $\omega = 0.0347$ rad / s
$\alpha = -\ 0.00241$ rad/s^2

6.50ps Actuator is extending at 0.00614 m/s^2

6.52ps

6.54ps $\mathbf{v}_{B'} = 1.15\,\mathbf{i} + 0.637\,\mathbf{j}$ ft/s

6.56ps $\mathbf{v}_{Arel} = -\ 15\,\mathbf{j}$ ft/s,
$\mathbf{a}_{Arel} = 45\,\mathbf{i} + 1\,\mathbf{j}$ ft/s

6.58ps $\mathbf{v}_{rel} = -33.7\,\mathbf{i} + 0\,\mathbf{j} + 17.5\,\mathbf{k}$ m/s
$\mathbf{a}_{rel} = -1.61\,\mathbf{i} + 0\,\mathbf{j} - 1.80\,\mathbf{k}$ m/s^2

6.60ps (a) $\mathbf{a}_A = -\ 2.58\,\mathbf{i} - 31.02\,\mathbf{j} - 3.86\,\mathbf{k}$ ft/s^2
(b) $\mathbf{a}_A = -\ 2.58\,\mathbf{i} - 30.91\,\mathbf{j} - 3.86\,\mathbf{k}$ ft/s^2

6.62ps ω_{max} is determined by edge speed
$\omega_{max} = 0.471$ rad/s = 23.9 °/ s

6.64ps $\mathbf{v}_{A/B} = -\ 1\,\mathbf{i} - 18.84\,\mathbf{j}$ m/s
$\mathbf{a}_{A/B} = -\ 29.6\,\mathbf{i} - 3.14\,\mathbf{j}$ m/s^2

6.66ps $\mathbf{a}_{S/A} = 7.23\,\mathbf{i} + 2.06\,\mathbf{j}$ ft/s^2

6.68ps arel = -0.000138 i + 0.000275 j + 0.000413 k m/s

6.70ps There are many answers to this problem, including stopping the object and returning to it on the next time around.

6.72ps w = 0.418 rad/s counterclockwise

6.74ps vArel = 8.73 i - 23.96 j m/s
aArel = - 0.836 i - 0.305 j m/s2

7.2ps ax = - 5/47 m/s^2
A = 1780 kN
B = 674 kN
friction = 1067 kN

7.4ps M = - 13.3 N m (clockwise)

7.6ps $\alpha = 3.76$ rad/s^2 (counterclockwise)

7.8ps $a_G = 2.97$ m/s^2
$\alpha = -\ 5.94$ rad/s^2

7.10ps Stopping distance = 7.53 m

7.12ps Rear wheel stays on ground
Stopping distance = 3.19 m

7.14ps $t_{stop} = 0.573$ s
$x_{stop} = 58.8$ ft
$N_A = 91500$ lb (main gear loading)
$N_B = 12500$ lb (nose gear loading)

7.16ps $a_{Gx} = -\ 4.91$ m/s^2
$N_A = 565$ N (rear wheel loading)
$N_B = 1005$ N (front wheel loading)

7.18ps (1) On Earth
(a) $\alpha = -\ 25$ rad/s^2

(b) $\alpha = - 18$ rad/s^2

(1) On Mars

(a) $\alpha = - 25$ rad/s^2

(b) $\alpha = - 7.38$ rad/s^2

7.20ps Increase weight of bar to 20 lbs gives $t_{stop} = 0.975$ s.
Increasing weight of disk to 25 lbs gives $t_{stop} = 1.22$ s

Thus, increasing the weight of the bar decreases stopping time while increasing the weight of the disk increases stopping time.

7.22ps $t_{hoop} = 1.89$ s
$t_{disk} = 1.65$ s
$t_{sphere} = 1.58$ s

7.24ps Top line is vGx and bottom line is w.

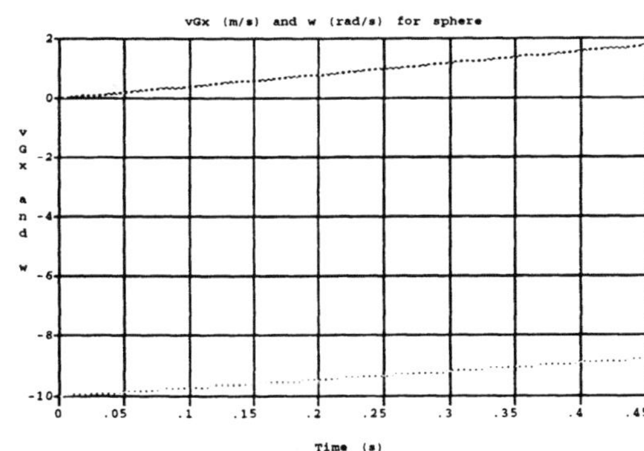

7.26ps $\alpha_{AB} = -12.71\ \mathbf{k}$ rad/s^2
$\alpha_{BC} = 25.42\ \mathbf{k}$ rad/s^2

7.28ps Disk is not homogeneous

7.30ps $I_{L'} = 0.88$ kg m^2

7.32ps $I_{L'} = 7\ m\ R^2 / 12$

7.34ps $I_{cm} = 25.84$ kg m^2

7.36ps $N_{nose} = 239.1$ lb
$x_{stop} = 463.7$ ft

7.38ps (a)$\omega(2\ s) = 769$ rad/s
(b) $\omega(2\ s) = 384.6$ rad/s
(c) $\omega(2\ s) = 192.3$ rad/s

7.40ps

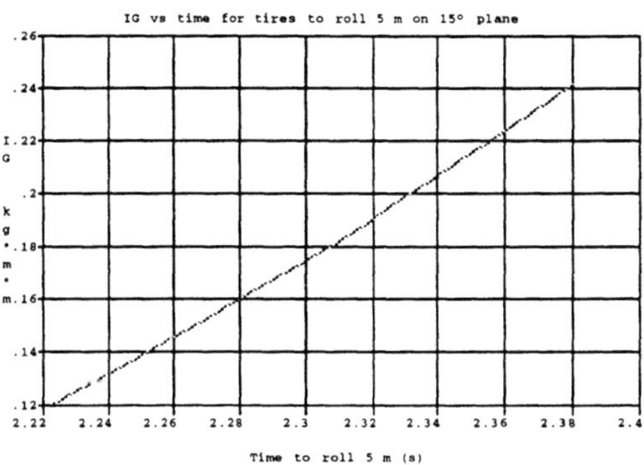

7.42ps $B_x = 12.08$ kN
$B_y = 7.083$ kN
$C_x = 12.27$ kN
$C_y = -6.609$ kN

8.2ps Work = 122,400 lb ft

8.4ps

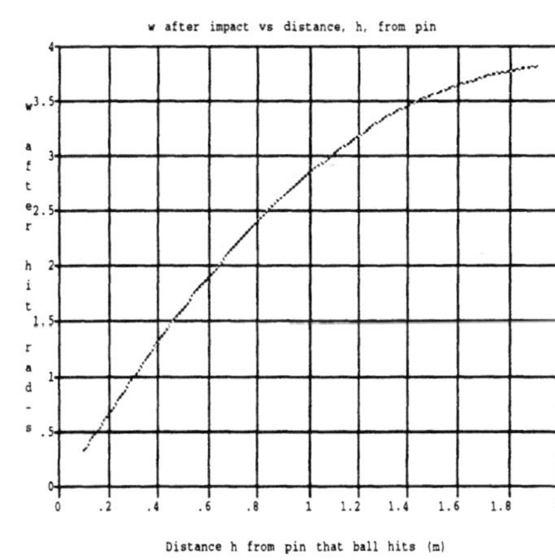

8.6ps $h_{max} = 1.134$ ft

8.8ps $v_1 = 1.286$ m/s
$v_2 = 1.714$ m/s
$\omega 1 = \omega 2 = 0.857$ rad/s

8.10ps $v_1 = 1.286$ m/s
$v_2 = 1.714$ m/s
$\omega 1 = \omega 2 = 0.857$ rad/s

8.12ps

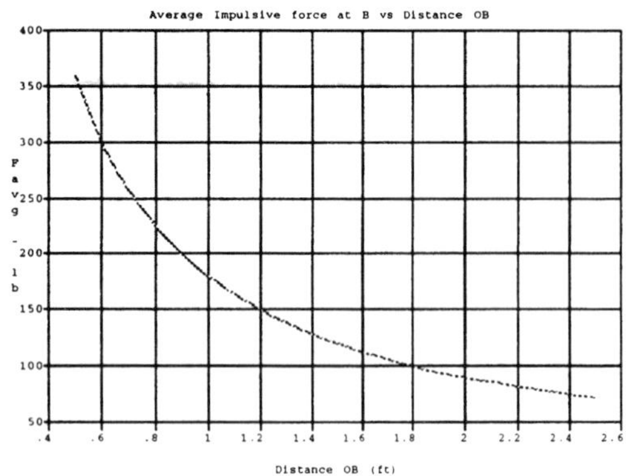

8.14ps

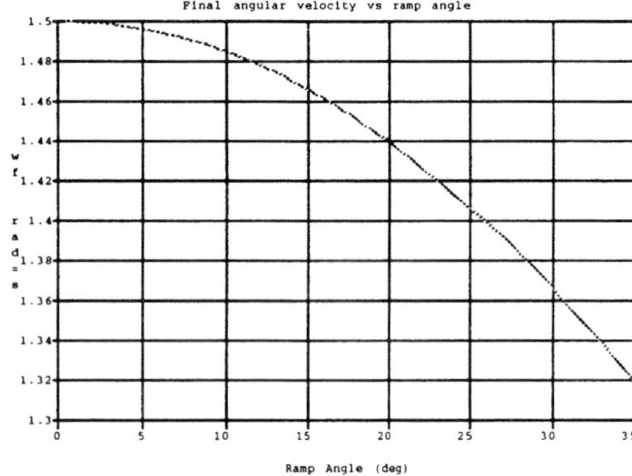

8.16ps

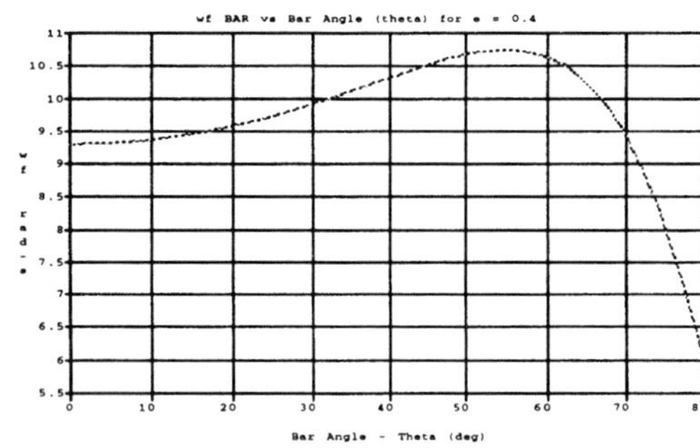

9.2ps $\mathbf{a}_{Q/P} = -\ 539\,\mathbf{i} + 320\,\mathbf{j} + 217\,\mathbf{k}$ m/s^2

$\mathbf{a}_Q = -\ 538\,\mathbf{i} + 323\,\mathbf{j} + 211\,\mathbf{k}$ m/s^2

9.4ps $\mathbf{a}_{B/A} = -\ 306\,\mathbf{i} - 1325\,\mathbf{j} - 31\,\mathbf{k}$ ft/s^2

$\mathbf{a}_B = -\ 304\,\mathbf{i} - 1316\,\mathbf{j} - 32\,\mathbf{k}$ ft/s^2

9.6ps $\mathbf{a}_L = -\ 206.2\,\mathbf{i} + 102.5\,\mathbf{j} + 42.5\,\mathbf{k}$ ft/s^2

$\mathbf{a}_R = 173.0\,\mathbf{i} - 89.9\,\mathbf{j} + 37.7\,\mathbf{k}$ ft/s^2

$\mathbf{a}_{L/R} = -\ 379.2\,\mathbf{i} + 192.4\,\mathbf{j} + 4.8\,\mathbf{k}$ ft/s^2

9.8ps $\mathbf{a}_C = 1050\,\mathbf{i} + 0\,\mathbf{j} - 950\,\mathbf{k}$ in/s^2

9.10ps $\mathbf{a}_P = 4.36\,\mathbf{i} + 10.89\,\mathbf{j} - 0.873\,\mathbf{k}$ m/s^2

9.12ps

The two maxima occur at 217° aand 323° (53° either side of the 270° point). The reason for the two maxima is that the dominant rotation in this case is the rotation w_o . The radius from the vertical axis is larger the farther away you go from the 90° - 270° line on the disk, and this causes the velocity of the point to be larger. However, the two velocities still interact as in 9.11ps.

v(217°) = v(323°) = 23.345 m/s

v(90°) = 17 m/s — the minimum.

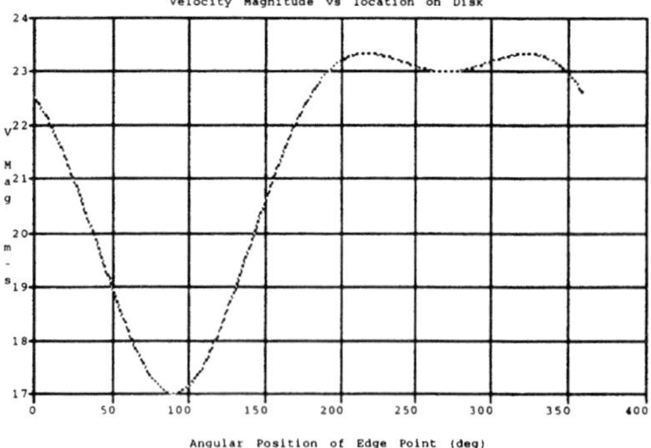

9.14ps

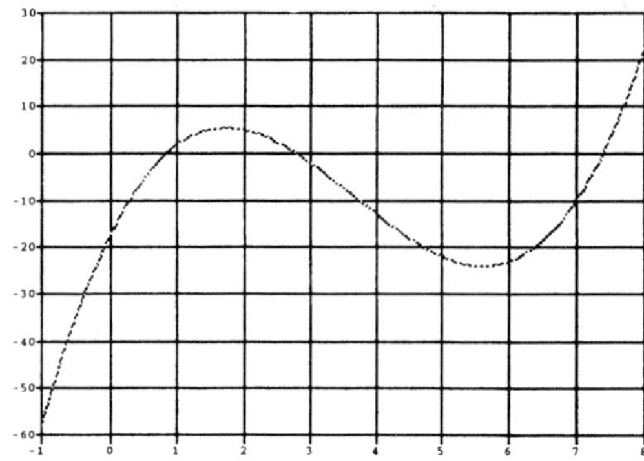

Roots are near 0.9, 2.75, and 7.4

These will be the guesses used. The converged values are

$I_1 = 0.824$ kg m^2
$I_2 = 2.79$ kg m^2
$I_3 = 7.38$ kg m^2

9.16ps $\mathbf{h} = 712\,\mathbf{i} + 179\,\mathbf{j} + 2362\,\mathbf{k}$ slug ft^2/s

9.18ps $I_1 = 22.32$ $\quad \varepsilon_x = 0.760\,\mathbf{i} + 0.608\,\mathbf{j} - 0.227\,\mathbf{k}$
$I_2 = 123.51$ $\quad \varepsilon_y = -0.649\,\mathbf{i} + 0.707\,\mathbf{j} - 0.279\,\mathbf{k}$
$I_3 = 177.17$ $\quad \varepsilon_z = 0.009\,\mathbf{i} - 0.360\,\mathbf{j} - 0.933\,\mathbf{k}$

9.20ps $I_1 = 0.358$ $\quad \varepsilon_x = 0.988\,\mathbf{i} - 0.152\,\mathbf{j} + 0.0\,\mathbf{k}$
$I_2 = 13.64$ $\quad \varepsilon_y = -0.152\,\mathbf{i} - 0.988\,\mathbf{j} + 0.0\,\mathbf{k}$
$I_3 = 14.00$ $\quad \varepsilon_z = 0.316\,\mathbf{i} + 0.949\mathbf{j} + 0.0\,\mathbf{k}$
Something is wrong here.

9.22ps $I_{xx1} = 2.48$ slug ft^2
$I_{yy1} = 5.59$ slug ft^2
$I_{zz1} = 8.07$ slug ft^2
$I_{xy1} = -1.86$ slug ft^2
The other products of inertia are zero

9.24ps $I_{xxcm} = 0.0398$ slug ft^2
$I_{yycm} = 0.0372$ slug ft^2
$I_{zzcm} = 0.770$kg slug ft^2
All products of inertia are zero

9.26ps $\varepsilon_x = 0.924\,\mathbf{i} + 0.383\,\mathbf{j} + 0.0\,\mathbf{k}$
$\varepsilon_y = -0.383\,\mathbf{i} - 0.924\,\mathbf{j} + 0.0\,\mathbf{k}$
$\varepsilon_z = 0.555\,\mathbf{i} - 0.832\,\mathbf{j} + 0.0\,\mathbf{k}$
Something is wrong here.

9.28ps $I_{xxcm} = 0.708$ kg m^2
$I_{yycm} = 5.875$ kg m^2
$I_{zzcm} = 5.5$ kg m^2
$I_{xycm} = 0$ kg m^2
$I_{xzcm} = 1$ kg m^2
$I_{yzcm} = -0.75$ kg m^2

9.30ps $I_1 = 0.308$
$I_2 = 0.643$
$I_3 = 5.048$

9.32ps Precession rate = 0.356 rad/s

10.2ps Natural frequencies are the same, the equilibrium position shifts.

10.4ps Period on Earth = 3.13 s
Period on the Moon = 4.14 s

10.6ps $x(t) = 0.1\,e^{-2t} + 0.2\,t\,e^{-2t}$

Notes

Notes

Notes

Notes

Notes

Notes

Notes

Notes

Notes

Notes

Notes

Notes

Notes

Notes

Notes